Management for Professionals

More information about this series at http://www.springer.com/series/10101

Jeremy David Curuksu

Data Driven

An Introduction to Management Consulting in the 21st Century

Jeremy David Curuksu
Amazon Web Services, Inc
New York, NY, USA

Amazon Web Services, Inc. is not affiliated with the writing and publication of this book nor to any material within

ISSN 2192-8096 ISSN 2192-810X (electronic)
Management for Professionals
ISBN 978-3-319-70228-5 ISBN 978-3-319-70229-2 (eBook)
https://doi.org/10.1007/978-3-319-70229-2

Library of Congress Control Number: 2017960847

Printed on acid-free paper

This Springer imprint is published by the registered company Springer International Publishing AG part of Springer Nature.
The registered company address is: Gewerbestrasse 11, 6330 Cham, Switzerland

Endorsements

"Jeremy David Curuksu's *Data Driven* is an extensive yet concise view on how the consulting industry will have to change in the age of data. His thoroughly researched perspective is accessible to a wide range of readers – also those without a consulting and/or data science background. However, this book is particularly relevant to anyone working in the management consulting industry, as it will shape our industry in the years to come."

–Christian Wasmer, Consultant, The Boston Consulting Group (BCG)

"Jeremy's book offers a comprehensive overview of the consulting industry–not only the type of cases and strategic approaches commonly used by the industry, but also applying these lenses on the industry itself to analyze the challenges it faces in the age of big data and where it might head next. In addition, it gives a great overview of the popular data science techniques applicable to consulting. The combination of the content will give consulting newcomers and veterans alike a new perspective on the industry and spark ideas on opportunities to combine the traditional consulting strengths and the new data science techniques in creative ways to offer distinctive value to clients. It'll also give anyone curious about what the black box of management consulting is about a good inside view on how the industry works."

–Yuanjian Carla Li, Associate, McKinsey & Company

"As consulting continues to go deeper into the business analytics space, this book provides great insights into both business frameworks and mathematical concepts to help you be a successful consultant. There is certainly a lot of material out in the marketplace. What I find to be most attractive about this book is that it doesn't matter if you are an experienced consultant like myself or just starting your career; there is lots of useful content for both. I personally found the statistical/analytical formulas as they pertain to business analytics paired with the business concepts to be extremely useful in getting a good understanding of the field. Highly recommended to both experienced consultants and those looking to enter this rapidly changing field."

–Thevuthasan Senthuran, Senior Strategy Consultant, IBM Corp., Chief Analytics Office

Acknowledgments

To the community of aspiring consultants at MIT, Annuschka Bork for her genuine suggestions, and my kind family.

Contents

Detailed Contents

List of Figures

List of Tables

About the Author

Jeremy David Curuksu The author is a data scientist, management consultant, and researcher. He worked at the strategy firm Innosight, at the Chief Analytics Office of IBM, and at Amazon Web Services. He holds a PhD in bioinformatics and was a research scientist for 6 years in applied mathematics at the Polytechnic School of Lausanne in Switzerland and at the Massachusetts Institute of Technology in the USA. He published 6 first-author peer-reviewed articles and presented 50+ seminars in data science and management consulting at MIT and at Harvard University. He is now advising Amazon's clients and working on solving their business issues through the use of artificial intelligence.

Introduction

Early in the twentieth century, a new trend started across decision-makers within most industries that responded to a burning desire to bring more discipline on what made good –or bad— *business sense*. This was the emergence of the *scientific method*, as applied to business and management. What Frederick Taylor then initiated is now referred to as the classical view of scientific management [1]. Indeed the science of management and strategic decision-making has carried on, evolving its own disciplines and sub-disciplines, its own principles, and what could at times be called theorems if it wasn't for the fear of uprising by mathematicians. It even evolved its own equations into business economics and game theory. But it seems that one would have had to wait until the twenty-first century, when increasingly complex technologies spread across all organizations down to the smallest family-owned businesses, to observe a more disruptive trend in the particular business of providing advice and assistance to corporate organizations: the industry of management consulting.

The emergence of new information technologies and what ensued, a *revolution* in the economics of information [2], have brought an unprecedented momentum to the application of scientific approaches in management consulting. The business of management consulting has always been about gathering data, analyzing it, and delivering recommendations based on insights gathered from the analyses. But nowadays, in an age of big data and computer-based analytics, the amount of "data" that one may gather in virtually every thinkable field of application, and the accompanying development of analytics capabilities, have increased by several orders of magnitude. Many IT firms started to specialize in data analysis and discovered natural applications in what used to be the exclusive domain of management consultants. In other words, the business of data analysis has become fundamentally different from what it used to be for at least as long as the management consulting industry ever existed (the first management company is believed to be Arthur D Little, which was created in 1886 [3]).

The potential disruption and future of artificial intelligence in the management consulting industry is the focus of Chap. 2 and discussed throughout the book. But for now and to the point, let us unveil the raison-d'être of this book: the management consulting industry, disrupted or not, will continue to demand scientific experts because the line where management consulting begins and computer-based analytics ends is becoming less and less evident. And computer-based analytics, with literally millions of available programs, requires some level of technical expertise indeed.

A second key factor is that even in the *standard* (i.e. not computer-based) model of consulting practices, grounded in deploying human capital and judgment-based solutions [4], a need for scientifically minded people also emerged. All major consulting firms have shifted their recruiting effort from MBA candidates toward a combination of MBA and non-MBA candidates –MD/PhD, scientists/experts, corporate managers. This evolving interest of consulting firms for scientists and executives is geared toward their subject matter expertise that, in addition to enabling actual technical expertise, has become essential to building trust with clients because corporations increasingly operate with complex new technologies. To some extent, this interest is also geared toward the analytic problem solving skills of scientists and corporate managers: everyone recognizes that it is easier to teach business to an analytically minded student than analytics to a business minded student.

Since many non-MBA scientists and professionals are unfamiliar with the business world yet legitimately seek to apply their scientific mindset within high-impact organizations, there should be a book that attempts to teach consulting (not just "interviews" as some easy-to-read volumes already do) without assuming prior business knowledge. This is a driving objective of this book, to offer a modern and complete introduction to management consulting for scientists, corporate managers and other professionals.

So in this book you will find a scientific introduction to management consulting. A fantastic book on this topic is *Management Consulting* from Milan Kubr [5]. The Kubr's volume is with no doubt one of the best books ever written on management consulting, but contains 950 pages and most of the text was written 20 years ago, yielding an obvious need for a more concise/modern volume. The book that you are reading has thus at least four objectives: to be scientific, modern, complete …and concise. It covers elementary and more advanced material, incorporates tools from data science, and discusses the emerging role of information technologies in consulting activities. The text draws on an extensive review of literature, with hundreds of peer-reviewed articles, reports, books and surveys cited, and on my personal experience as a consultant and a scientist. It is also supported by insights gathered at volunteer cross-company workshops and introductory lectures that I coordinated during 2013–2015 as the leader of the consulting community at MIT.

A website accompanies this book and is accessible at *econsultingdata.com* without additional charges. The *econsultingdata.com* website is a resources platform for management consultants. It can also be accessed from the MIT consulting club's website and a few other partner websites. It redirects toward books, articles, tutorials, reports and consulting events. It also contains original contents that represent an interactive version of some parts of the book (e.g. industry snapshots of Chap. 2 and toolbox of Chap. 4).

This book assumes no technical background. It introduces consulting activities and demand determinants in key markets, delves into the client-consultant relationship, discusses the interface between data science and management consulting, and presents both primer and advanced material in data science and strategy. An entire chapter is dedicated to outlooks in the industry. Covering this breadth of topics in one book has not been attempted before, which is not surprising since big data (computer-based analytics) flourished only a few years ago. But today big data is everywhere, and some articles have started to discuss how AI is disrupting the consulting industry [4, 6]. What was considered technical 10 years ago is now common knowledge. Or so it owes to be.

To deliver a complete introduction to management consulting in the twenty-first century the book had to include at least one chapter on data science yet assume no background in statistics. Given the challenge that this objective represents, the introduction to data science was split in two chapters: *Primer* and *Advanced*. Beginners might not easily follow the *Advanced* chapter, but they are encouraged to persist! To help with that every concept is introduced using popular words, down to the notion of p-value. In this sense, here is a unique opportunity for *anyone* to overview the maths behind *data science* and thereby what may soon become the fundamentals of management consulting in the twenty-first century.

The disruptive impact of new information technologies on management consulting activities is thoroughly incorporated in this book: Chap. 1 analyzes the entire industry as of early 2017, presenting the value proposition, the different market segments, market players, and success factors. The impact of new information technologies naturally emerges throughout. Three chapters are then entirely dedicated to this modern dimension of management consulting: Chap. 2 (*Future of Big Data in Management Consulting*) and Chap. 6 and 7 (*Principles of Data Science*). In Chap. 2 a scenario planning exercise is presented to help the reader frame the possible future scope of big data in management consulting. In Chap. 3 (*Toolbox of Consulting Methods)* all fundamental categories of consulting activities are overviewed. Traditional consulting activities are described and augmented by computer-enabled ones. For each category a concise "recipe"-like method is proposed.

The aspect of client interaction (Chap. 4) is often bypassed in existing books on management consulting, yet everyone agrees that it is an essential difference when comparing management consulting to other professions in management, for example corporate management or academic research. And thousands of articles have been written about this topic. In this chapter, we will thus clarify "what" is the nature of the relationship between a consultant and his/her client. Different types of relationships have been articulated in the literature, and it is essential to understand how the nature of the relationship, the client's *perception* of what represents success in a consultancy, and the actual success of a project are all connected. This chapter ends with key insights and takeaways from the literature pertaining to client expectations and interactions along the different phases of a project.

Chapters 5, 6, 7, 8, and 9 discuss the tools, methods and concepts that enable a consultant to inquire, diagnose and eventually solve a client's problem. Chapter 5 (*The Structure of Consulting Cases*) builds upon the popular process of inductive reasoning described in Chap. 4 to help the reader tailor high level thinking roadmaps and effectively address problem statements (in simulation case-interviews as in real life…). For this purpose, a set of both generic and not-so-generic MECE frameworks are given, and most importantly this dilemma (between *one size fits all* a.k.a. *taken off the shelve* and *reinventing the wheel* a.k.a. *built from scratch*) is addressed head-on and explicitly explored. This same dilemma also exists in the vast landscape of tools and approaches that consultants use everyday. The merit of defining categories of tools and approaches may be purely pedagogic. Of course, developing a strategy should always involve holistic approaches that appreciate the many potential interactions and intricacies within and beyond any proposed set of activities. But again, one has to start somewhere: Chaps. 8 and 9 describe the many types of concepts, tools and approaches in the field of strategy. These chapters categorize and overview best practices, and place special attention on practical matters such as key challenges and *programs* that can be used as roadmaps for implementation.

As for the entire book. The reader is invited to consider these facts, models, tools and suggestions as a simple aid to thinking about reality. No concept or serious author thereof has ever claimed to outline the reality for any one particular circumstance. They claim to facilitate discussion and creativity over a wide range of concrete issues.

Analysis of the Management Consulting Industry

1

In this introductory chapter, the management consulting value proposition is put into context by looking at general trends and definitions across different segments, key success factors, competitive landscape and operational value chain.

1.1 Definition and Market Segments

After a brief overview of the value proposition and industry life cycle, different segments are defined in turn by services (Sect. 1.1.3), sectors (Sect. 1.1.4) and geographies (Sect. 1.1.5) to look at demand determinants from different perspectives.

1.1.1 The Value Proposition

Management consultants provide advice and assistance to organizations within strategic, organizational and/or operational context [7–10]. An elegant definition was given 30 years ago [11], and is still totally adequate today:

> *"Management consulting is an advisory service contracted for and provided to organizations by specially trained and qualified persons who assist, in an objective and independent manner, the client organization to identify management problems, analyze such problems, and help, when requested, in the implementation of solutions"*
> Greiner and Metzger 1983

Why are business organizations purchasing management consulting services? A universal answer to this question cannot exist given the vast range of services offered in this industry which caters boundlessly from strategic planning, financial management and human resource policies to process design and implementation, just to name a few. Management consultants are generically considered *agents of change* [7] whose value proposition relates both to functional and cultural (a.k.a. psychological) needs of their clients. The nature of the client-consultant

J. D. Curuksu, *Data Driven*, Management for Professionals,
https://doi.org/10.1007/978-3-319-70229-2_1

relationship will be discussed in Chap. 4. In this chapter we focus on facts and figures that will help us understand the industry as a whole.

1.1.2 Industry Life Cycle

As of 2017, the industry was in the growth stage of its life cycle with an annualized 3.4% increase rate in the number of firms expected within the next 5 years (to ~1 M) and an annualized 3.6% growth in revenue (to $424bn) [9]. This came largely as a result of the industry's continued expansion into BRIC countries and most specifically China and India. Increased demand for management consulting in emerging economies is significantly boosting the industry and expected to continue over the next few years.

The rapidly growing market demand in emerging economies has helped offset the impact of the global recession in developed economies. In the United States and Europe the recovery is expected to be gradual and prolonged. The impact of the global economic downturn on the management consulting industry has been moderate compared to most other industries. Revenue declined by 2.4% in 2009, 3.5% in 2010, and gradually recovered toward positive figures thereafter. Interestingly indeed, this industry benefits from a counter-cyclical demand for its services: a base level of demand is ensured in times of economic downturns because consultants can assist clients mitigate their losses; in times of economic prosperity or recovery consultants can assist clients develop more aggressive profit-maximizing strategies.

At least two current trends might be noted in the consulting industry life cycle. First, consolidation is on the rise in the United States and Europe. This contributes to expanding the industry's service offerings (through M&A between large firms and smaller specialist firms) and may reflect a locally saturated market that reached maturity as hypothesized by IBIS World [9]. To provide an alternative hypothesis though, it may as well reflect a temporary response to business uncertainties and volatile financial markets that have accompanied a slow recovery from the recent economic downturn.

Second, the industry is evolving toward a broader scope of overall service portfolio. In particular, the distinction between management consulting services and more technical IT consulting services is becoming less and less evident [4, 13], thanks to new business "computer-based" possibilities offered by the phenomenon of *big data* (described in Chap. 2).

1.1.3 Segmentation by Services

1. 40% – *Business strategy*

This segment is covered by most management consulting firms, generalists and specialists alike. It involves the development of an organization's overall business direction and objectives, and assists executive decision-making in all matters such as growth, innovation, new ventures, M&A, outsourcing, divestiture and pricing (see Chaps. 8 and 9). Both academic researchers as well as corporate organizations contributed to the establishment of the tools and concepts that are now widely used in management strategies [3], for example Peter Drucker from NYU/Claremont [14] and Bruce Henderson from BCG [15]. Chapter 8 will cover the basic concepts and tools in strategy. Chapter 9 will discuss some more advanced concepts.

2. 15% – *Marketing management*

In this segment, management consultants assist their clients with positioning, pricing, advertising, attracting new customers, developing new markets and bolstering brand awareness.

Some simple marketing frameworks have been developed [16] that capture the grasp of the breath and depth that often come with marketing consulting projects. For example, the "*4Ps*": Product (positioning, fit, differentiation, life cycle), Promotion (advertising, promotional campaigns, direct/personal sales campaign, public relation), Place (exclusive, selective, mass distribution), Price (cost-based, competition-based, customer value-based, elasticity). And the 5Cs (Company, Competitors, Customers, Collaborators, Climate). Both frameworks will be detailed in Chap. 8.

3. 10% – *Operations and value chain management*

The successive phases that a product or service goes through from raw material/information supply to final delivery is often referred to as the *value chain* [17]. Going beyond the organization as a unit of reference, Harvard Professor Clayton Christensen further developed the concept of *value network* [18] where an organization's value proposition and business model fit into a nested system of producers and markets that extends up into the supply chain and down into the consumption chain. The concept of value network comes with the elegantly embedded concept of *jobs-to-be-done* [19] that customers hire products or services to do. The distinction of this consulting segment, the management of value chain/network, with other segments/activities is thus not obvious. Depending on the circumstances, it might include specialization strategies (e.g. cost-reduction, differentiation, or focus), quality system management, inventory management, scheduling, warehousing, and even entire business model re-design.

4. 10% – *Financial management*

This segment includes services in banking, insurance and wealth management (securities distribution, equity investment, capital structures, mutual funds, etc). The nature of the services provided by external consultants tends to be less and less related to financial management itself and more and more related to other consulting services (e.g. strategy, operation, marketing) because most large financial institutions have internalized their own financial management counseling services.

Nonetheless, the actual services in vogue with these clients (strategy, operation, marketing) require external consultants to develop some basic knowledge in financial management (e.g. option pricing, portfolio theory). Section 7.3 and Chap. 8 will discuss some basic concepts in finance.

5. 10% – *Human resource management*

This segment includes services in human resource policy development, process design, employee benefit packages and compensation systems, etc. As for financial services, many large organizations have internalized these services or else contract specialist firms that focus exclusively on human resource management systems.

6. 15% – *Others*

The consultant's role is uniquely versatile in its nature. The "miscellaneous" segment thereby includes many types of projects, which, if they may be clearly defined, do not fall into any of the "typical" categories described above, e.g. some projects that focus on accounting, governmental programs, or technology development.

There are also projects for which the initial problem (i.e. when the project starts) may span across any number of categories. For instance, an issue related to innovation might involve a multitude of potential solutions, some of which may relate to marketing, others to new technology, yet others to business model re-design. Depending on market circumstances, client capabilities and other factors, two projects starting with a similar question may evolve toward completely different directions.

1.1.4 Segmentation by Sectors

In this section, nine sectors were selected that represent the vast majority of industries in which management consultants offer services. A concise introduction to each of these industries is given, by reviewing in turn its products, distribution channels, customers, competitors, revenue streams, cost structures and overall market trends. For quick references, this section can be accessed in an interactive format at econsultingdata.com, the book's accompanying e-platform.

1. *Healthcare and pharmaceutical*

Biopharmaceutical segment [20–23]	
Product	Biopharmaceutical products include patent-protected and generic drugs that can be obtained either by prescription or over-the-counter, and target either human or animal diseases
Distribution	Prescription drugs: Pharmacies, hospitals/clinics, B2B OTC drugs: Retail outlets, pharmacies, email orders, B2B
Customers	Health care providers, payers (e.g. HCO, insurance companies), patients, pharmacies, hospitals, and government in some emerging markets
Competition	Product quality (efficacy, safety, convenience), brand name and control of distribution are the major basis for competition Price competition from generic manufacturers increasing
Key trends	1. Major treatment areas are: oncology, psycho/neurology and cardiovascular 2. R&D challenge is to find high revenue blockbuster drugs 3. Price competition from generic manufacturers 4. Pressure from government and payers to decrease prices 5. High risk of not getting approval from regulatory bodies (i.e. high attrition rate) 6. Emerging markets growing notably for outsourcing 7. Demographic shift: aging population
Revenues	Key revenue drivers include the size of specific treatment domains, buy-in from doctors (best-in-class), speed to market (first-in-class), level of competition, expertise for formulation of generics and networking/advertising
Costs	High cost for R&D including discovery, formulation and clinical trials, for manufacturing (economy of scale) and for marketing (sales, promotion), which are key barriers to entry

Care centers and hospitals [24–26]	
Product	Surgical and nonsurgical diagnostics, treatments and operating services for in- and out-patients
Distribution	Direct through personnel (practitioners, nurses)
Customers	In- and out-patients with medical condition, payers/insurance companies
Competition	Quality of care, breadth of the service portfolio and skilled workforce are the major basis for competition. Increasing competition from specialty care centers
Key trends	1. Shift toward outpatient care models 2. Shift from a fee-for-service to a value-based (i.e. outcome-based) payment model 3. Pressure from government and payers to decrease prices 4. Governmental policy changes e.g. Affordable Care Act (*Obamacare*) in the US 5. Demographic shift: aging population
Revenues	Some key drivers are access to highly skilled labor, proximity to key markets, reputation, optimum capacity utilization, and understanding of government policies
Costs	Main costs incur wages, marketing, purchase of medical equipment and pharmaceutical supplies

2. *Financial services*

Consumer banking segment [27, 28]	
Product	Credit cards, consumer loans, deposit-based services, securities/proprietary trading
Distribution	ATM, Online, branches/tellers
Customers	Individuals, *high net worth* customers, small/medium businesses without financial service
Competition	Large national players and regional banks
Key trends	1. Increasing use of ATM and online distribution 2. Demographic shift: aging population 3. Increasing offshoring of call centers and back office functions 4. Primary growth through M&A
Revenues	Fees, borrowing rates
Costs	Borrowing costs, overhead (branches, administrative, compliances), salaries, *bad* debts

Private equity/investment banking segment [29–31]	
Product	Securities, venture capital, growth capital, mezzanine capital, leveraged buy-outs, distress investments, …
Distribution	Direct through personnel, mutual funds
Customers	Small family-owned companies, large corporations, institutional investors
Competition	All sizes of PE firms compete with each other
Key trends	1. The business revolves around *go* versus *no go* investment decisions 2. Number of deals in decline 3. Deals tend to involve larger amount of cash: customers tend to be larger corporations
Revenues	Return-on-investment/time horizon, with a strong dependence on financial (access to capital, capital structure) and operational (more efficiencies, new management) levers that may be pulled
Costs	Since major costs are the funds required to invest, these represent opportunity costs

3. *Insurance*

Insurance segment [32]	
Product	Liability for various types of risks (car crash, fire damage, credit default)
Distribution	Sales force and online sales
Customers	Individuals and all types of businesses
Competition	Niche players and large players operating across multiple segments
Key trends	1. Marketing through better websites easier to use 2. Governmental policy changes such as *Obamacare* are regularly changing the landscape of insurance markets
Revenues	Premium collected; revenues heavily depend on managing risks and controlling costs
Costs	Claims (payments), overheads (administrative), salaries, sale commissions, marketing

4. *Media*

Media segment [33, 34]	
Product	Generation and dissemination of audio/video contents and printed media. Consumers are part of the product in the traditional business model (advertising-based revenue model, see below)
Distribution	For printed medias: papers, online and mobile For TV-A/V: traditional broadcast/cable, online and mobile For movies: theaters, rentals, online and mobile
Customers	In the advertising-based revenue model: advertisers In the subscription-based revenue model: consumers
Competition	Both in the advertising- and subscription-based models the audience interest is the basis for competition
Key trends	1. Consumers are part of the product 2. Emergence and increase of business models based on subscription due to internet 3. Digitalization harmed as well as created new opportunities in the media sector
Revenues	Advertising and/or subscriptions
Costs	Fixed costs: studios, printing presses, overheads, new technologies Variable costs: marketing, salaries

5. *Telecommunication and information technology*

Telecommunication and information technology segment [35–37]	
Product	Hardwares (servers, PCs, semiconductors, communication equipment) Softwares (algorithms, IT services) Internet (search engines, portals)
Distribution	Direct carrier-owned brick-and-mortar/online stores Indirect retailer-owned brick-and-mortar/online stores
Customers	Consumers, B2B, retail outlets and government
Competition	High competition between large multinational corporations has led to a phenomenon of *coexistence,* i.e. a collaborative ecosystem between competitors
Key trends	1. High consolidation through mergers and acquisitions 2. Coexistence (see above) fosters one-stop-shops 3. Cloud computing makes it increasingly easy for corporate customers to outsource their IT operations, which benefits the industry 4. The spectacular growth of the mobile phone penetration over the past 20 years is expected to continue and reach 80% globally by 2020 [38]
Revenues	Software: license/maintenance model or subscription revenue model Internet: revenue per click and advertising-based model; Telecom/mobile: advertising, subscriptions, data services, app. stores
Costs	Fixed costs: R&D, equipment, staff utilization, overheads, infrastructures Variable costs: marketing, salaries

6. *Consumer products*

Consumer products segment [39]	
Product	Household products (e.g. soap, snack, food, pet supply)
Distribution	Retail outlets, wholesalers, direct email order/online sales
Customers	Individual customers, retailers of all sizes, wholesalers
Competition	Product mix and brand management are the basis for competition. New products and innovation are critical to success
Key trends	1. Lifestyle and behavior of consumers drive demand, e.g. aging population, online advertising, social network, economic downturn, go-green 2. Perpetual extension of product line from both internal and outside-in innovation 3. Increasing influence of governmental regulations
Revenues	Low margin products with revenue based on volume High margin branded goods that foster price premium
Costs	COGS (e.g. raw material, packaging), sales, marketing, branding

7. *Manufacturing*

Manufacturing segment [40]	
Product	Mechanical, physical or chemical transformation of materials or substances into new products (e.g. textile, electronic equipment, chemicals, machinery, vehicles)
Distribution	B2B, retail outlets, wholesalers
Customers	Corporate customers, individuals, government
Competition	Management of supply chain, process efficiency and distribution are a main basis for competition
Key trends	1. New technologies, adjacent industries and emerging markets drive demand 2. Manufacturing sectors are highly cyclical 3. Increase of outsourcing to low-cost emerging parts of the world
Revenues	Low margin products (e.g. automobile) with revenue based on volume High margin products (e.g. aircraft) that foster price premium
Costs	Capital investment, raw material, labor, marketing

8. *Energy and utilities*

Energy and utilities segment [41, 42]	
Product	Production and supply of energy and utilities. Notably: exploration, development and operation of oil and gas fields
Distribution	B2B
Customers	Petrochemical companies, natural gas distributors, electricity generators
Competition	Price is a key basis for competition but because the energy market is commoditized and volatile, the control of distribution networks based on long-term contracts is essential to competition
Key trends	1. Volatile gains due to occasional shortfalls and regulatory frameworks 2. Increase of trade-based international exchanges, which reached 50% of US oil consumption in 2016 3. Advanced economies represent the majority of the market but BRICs drive growth 4. High consolidation through mergers and acquisitions 5. Increase of competition from alternative sources (wind power, coal, etc)
Revenues	Revenues are driven by long-term distribution contracts, the ability to find new resource deposits and the ability to comply with regulatory frameworks
Costs	Price of raw material, equipment, marketing, R&D

9. *Airline*

Airlines segment [43, 44]	
Product	Air transport services for freight and individual passengers
Distribution	Direct by telephone, Internet, OTC/walk-ups, travel agents
Customers	Corporations, small businesses, individual customers, travel agents/websites
Competition	Price is the major basis for competition
Key trends	1. Increasing consolidation through mergers and acquisitions 2. Increasing competition from low-cost carriers
Revenues	Load-based fees, individual tickets, baggage fees, food and beverages
Costs	Fixed costs: Aircraft, airport gates, labor, IT and administrative Variable costs: Fuel, hourly employees, food/beverages

10. *Others*

Demand from a particular sector may be strongly affected by factors such as overall corporate profit, business confidence, government investment decisions, and global economic cycles. It can also change substantially due to the introduction of new products, technologies or government policies. While the list above is not exhaustive, many of the trends and models described may be used as potential guides for omitted sectors or sub-sectors. In addition, they may be used as ideation starters for potentially fruitful transfers of ideas from one industry to another.

Governments in themselves represent a significant segment, but IBIS World [9] reports that expenditure on consultancy services from governments is increasingly becoming a contentious political issue for lack of tangible assessment of the benefits obtained from these services. For example in the US, governments at federal and state levels handle highly rigid budgets and seek high-impact outcomes from which voters can immediately benefit. But the need for short-term tactics is often not amenable to the strategy-based long-term benefits that management consultants are trained to deliver.

1.1.5 Segmentation by Geography

The global management industry is concentrated in North America and Western Europe but BRIC economies are expected to drive growth over the next 5 years [9]. The geographical proximity to key sectors (e.g. finance, biopharmaceutical), the access to highly skilled workforce and the ability to provide in-person client assistance, are all essentials to the success of a management consulting firm. Not surprisingly, these factors drive the nature of international expansion. For example, the predicted growth in China and India correlates with the rapid expansion of a tertiary service-based culture in these two countries.

1. 50% – *North America*

The United States is still by far the largest market for management consulting, where historically most large corporations saw the light of day [3], e.g. Arthur D. Little, McKinsey & Co, The Boston Consulting Group and Bain & Co. Accounting and audit firms such as Deloitte and PwC increasingly refocus on management consulting services; it now represent the majority of their income [10] in the United States! Industry concentration revolves around New York, California, Massachusetts and Pennsylvania.

2. 30% – *Europe*

The high level if interconnectedness fostered by a solid establishment of the European Union over the past 20 years have facilitated corporate expansion beyond national borders, and greatly contributed to increasing the demand for management consulting services. Some cultural differences with the US model are apparent. For example, most new recruits in European firms and local offices come with MBA-like or professional training, while scientific PhD-trained students now represent 1/3 of new recruits at McKinsey and BCG's North America offices [45].

Industry concentration revolves around the largest cities: London, Berlin, Paris and Zurich. The sovereign debt crisis has hindered growth in overall Europe, mostly as a result of low business confidence and unstable demand in countries such as Greece, Ireland, Spain and Portugal.

3. 10% – *India and China*

The economic growth and evolution toward a more service-based culture in India and China fostered an exponential expansion of local businesses and the implementation of many global corporations in these regions. As a result, the demand for management consulting and in-person services in these regions has increased dramatically.

China leads the way so far but India offers peculiar opportunities thanks to its different political environment. For example, outsourcing and offshoring to India is more popular than to China. A large body of research on innovation [18, 46, 47] underscores the importance of appreciating the specific needs of emerging markets and tailoring innovative business models to these *jobs-to-be-done*, rather than designing a model on the assumption that because it is very successful in other regions of the world, it should there too. With the right tailored model, who knows how successful a company might become with a growing target market of several billion consumers?

4. 10% – *Others*

South America, Russia and other regions in Central Asia are growing rapidly too. As for India and China, the demand for management consulting and in-person services in these regions has increased dramatically. Regions such as Brazil, Mexico, Peru and Chile are currently emerging economies whose fate concur with the US economy because of their exportation models.

Finally, South Africa and the Middle East are growing relatively slowly, which researchers often attribute to poor political environment and civic turmoil. The demand for management consulting and in-person services in these regions increases proportionately.

1.2 Success Factors[1]

1. *Employees productivity is at the core of company performance*

The *product* in a management consulting firm is just as good as the individuals it employs. Wages represent 60% of cost structures on average. Company executives in this industry have an exceptional talent for innovating with employee productivity and quality control systems. These innovations often emerge from the frequent *mutual feedbacks* between managers and staff, team building exercises, in-house tools and cross-project learning. A productive relationship between its team members is essential to the success of a consulting company.

2. *Access to highly skilled workforce*

Again, the product is the consultant. Management consulting is knowledge-based and skill-intensive, it requires a workforce comfortable with analytics and teamwork. Most global corporations have implemented campus-wide recruiting campaigns that follow the academic cycle around top universities across the world.

3. *Focused services and corporate image are the basis of competition*

From the client's perspective, it is impossible to know with certitude whether he/she contracted the consulting firm that was the best fit for tackling his/her issue. *The client cannot know in advance what she is exactly buying before she gets it* [48]. Hence, to be competitive, a consulting firm needs to propose a clear differentiator or price advantage. Differentiators often take the form of referrals from past assignment, brand name and sharp focus on a service (e.g. strategy) or sector (e.g. healthcare).

4. *Long term client relationships are essential to sustaining growth*

Repeat assignments are needed to effectively market consulting services and grow a reputation. It is almost always easier to maintain an on-going relationship with a client than to develop a new one (see Chap. 4). Here again, effective internal

[1] Adapted from the 2016 IBISWorld report on Global Management Consulting [9].

processes are keys. So-called *Partners* (and/or *Principals*) in consulting firms are the agents whose role is to initiate and maintain client relationships, and convince a prospective client that quality output will be delivered on time and on budget.

5. *Small- and medium-size firms successfully leverage access to Niche markets*

As for many other unrelated industries, economies of scale may dictate the scope of product portfolio. In management consulting, this implies that new ventures will build a most effective business only if they define a sharp focus on a well-defined target market. Many examples are available of companies that became very successful by leveraging a clear focus on a given sector (e.g. pharmaceutical, finance, aerospace) or a given service (e.g. design, innovation, marketing).

1.3 Competitive Landscape

1.3.1 Basis of Competition

In the management consulting industry the basis for competition remains heavily focused on differentiation based on quality service such as knowledge, skills and expertise [9]. Client satisfaction and reputation represent a fundamental competitive edge in this industry, where referrals and testimonials from past clients are the main drivers of repeat customers and client base expansion.

Price-based competition tends to be leveraged during low economic growth periods, and more generally by the smallest players such as freelancers and organizations with highly specific offerings such as technical consulting services, HR, accounting and international laws.

Awareness of local cultural issues and business practices is an all too often-overlooked competitive lever [46]. For example, the innovation opportunities in advanced economies versus emerging ones have different characteristics and require different business models and product specifications [18, 46]. This has become particularly important today as the consulting industry growth is expected to largely comes from expansion into BRIC markets. Global consulting corporations are increasingly deploying offices in developing countries, most particularly Asia.

The range of services offered in management consulting is expanding. There are clear trends toward more and more implementation of the recommendations, and more and more integration of IT based services [4]. These expansions are enabled either by developing new internal capabilities (e.g. the *McKinsey Solutions* analytics platform) or strategic alliances with specialist firms.

1.3.2 Emergence of New Information Technologies

Computer-based information technologies have become increasingly relevant to management, production and logistics strategies [2, 49]. This trend represents both an opportunity for innovation and a threat to management consulting firms. The scenario planning analysis developed in the next chapter will articulate the outlooks that new information technologies present in management consulting from either perspective – opportunities and threats – and map out potential scenarios.

Several examples illustrate the disruptive movement of computer-based data analytics into formerly judgment-based management consulting activities. Many start-ups recently emerged in the arena of big data analytics with the sole mission of assisting corporate organizations improve their business, and most large global service companies have already revisited their portfolio in this direction. IBM moved toward management consulting by creating a "*Global Business Services*" division; Booz Allen Hamilton moved toward data analytics by creating a "*NextGen Analytics*" division. Meanwhile at McKinsey and BCG, the "*Solutions*" and "*Gamma*" units respectively offer to equip clients with customized analytics tools.

In certain circumstances, IT consulting is beginning to be seen as a viable alternative to management consulting [4]. The distinction between management services and more technical IT services indeed is becoming less and less evident [2], thanks to the new possibilities offered by the phenomenon of *big data.* More details are presented in Chap. 2.

1.3.3 Main Players

The management consulting industry is a fragmented market (i.e. low concentration). The top seven players represented less than 15% of global revenue in 2016 [9]. This is usually explained by the broad range of services offered and because the industry is still in a growing phase [9]. The global market is expected to become even more fragmented over the next few years due to emerging economies. Consolidation is expected to rise marginally and exclusively in advanced economies.

To first approximation, three categories of organizations can be considered: the global mega-firms (so-called *Top Three* and *Big Four,* recently joined by IT conglomerates such as IBM), the specialist firms that have developed an intellectual property and a reputation around a given set of services and industries, and the myriad of emerging startups and freelancer operators contracted for niche requests. Anyone can call oneself a consultant indeed! If you have an extended LinkedIn network, you will notice that many job seekers indicate *consultant* on their profile for lack of a better title. This creates confusions when discussing the competitive landscape of management consulting. Note thus that in Fig. 1.1, the figures refer to formally registered companies.

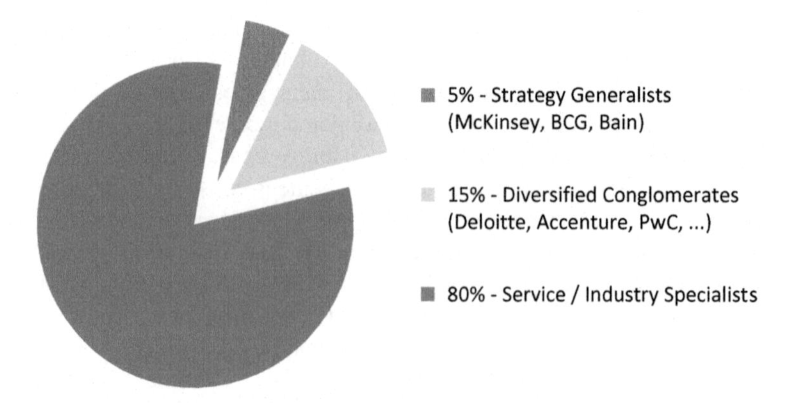

Fig. 1.1 Aggregate 2017 market share of generalists, conglomerates and specialists

1.4 Operations and Value Network

This section provides a brief overview of the value proposition in the context of its *value network*[2] and key activities in management consulting. These activities are discussed in more depth in Chap. 3 and across the book. Note that neither this section nor Chap. 3 delve into the different chronological phases of an assignment, because this has more to do with developing a client-consultant relationship than developing operational capabilities. The client-consultant relationship is the focus of Chap. 4, and for this reason the chronological phases of an assignment will be covered in Chap. 4 (Sects. 4.4, 4.5 and 4.6).

A customized service Management consultants serve a distinctly personalized function. One consequence is that the client cannot know in advance what she is exactly buying before she gets it. Conversely, she may expect from her consultant to dedicate and appreciate the peculiar intricacies of the problem at hand and develop a tailored solution. Management consultants are problem-solvers; agents of change [7] whose value proposition responds both to functional and cultural (a.k.a. psychological) needs from their clients (Fig. 1.2). So this is what the client pays for.

Supply The industry is labor-intensive [8–12, 49]. Because consultants are the primary assets of a consulting firm, the firms invest considerable amounts on formal training, mentoring and professional development programs (Fig. 1.2). They also invest in the purchase of information technologies: databases, libraries and intranet systems that hold reports from past assignments and enable a knowledge-based intellectual property to build up.

[2]The concept of value network [18] is described in Sect. 1.1 and Chap. 7.

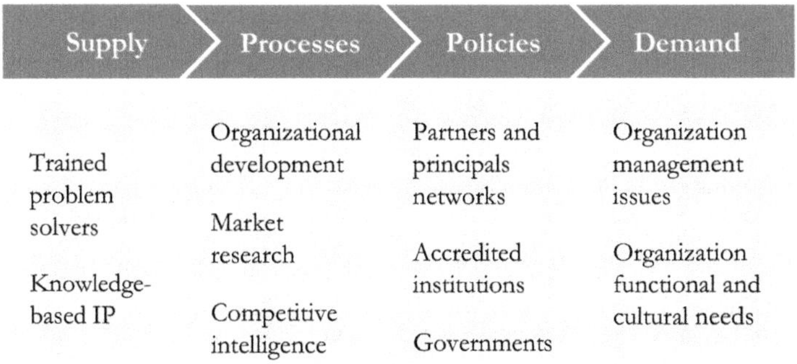

Supply	Processes	Policies	Demand
Trained problem solvers	Organizational development	Partners and principals networks	Organization management issues
Knowledge-based IP	Market research	Accredited institutions	Organization functional and cultural needs
	Competitive intelligence	Governments	

Fig. 1.2 The management consulting value network

Demand Depending on the scope of the organization, new clients may be sourced through domestic or international networks. So-called *Partners* (and/or *Principals*) in consulting firms are the agents whose role is to initiate and maintain client relationships. The client relationship (discussed in Chap. 4) involves a high degree of personal interaction with the client including daily communications and frequent visits on-site.

Policies The management consulting industry is seldom regulated by governmental bodies, except in developing countries where governments frequently tamper with the decision-makings of local corporate organizations [9]. In advanced economies, policies applied to management consulting may come in the form of accreditation to recognized institutions such as the AMCF [50], but the type of credibility bolstered through proprietary brand development and personal networking by partners and principals is the most common and effective form of accreditation (Fig. 1.2). Personal networks substitute to the type of credibility offered by formal accreditation. Still, given that 45% of management consultants are freelancers or newborn startups [10], an AMCF's *license to operate* can give a significant push to recent nascent consulting ventures.

Processes Consulting activities may be organized in three categories – organizational development (OD), consumer/stakeholder market research (CMR) and competitive intelligence (CI). The most common activities are listed in Table 1.1. The first category regards issues internal to the client organization, the second issues with external stakeholders and the third issues with market players (who may represent competitors, partners, or acquisition targets). These three categories will be used to organize the discussion in Chap. 3 where each activity is introduced with an itemized action-plan. Of course the merit of defining such categories is purely pedagogic, since any activity may be adapted to fit the need of a particular case and

Table 1.1 Sample of consulting activities: organizational development (OD), consumer/stakeholder market research (CMR) and competitive intelligence (CI). Each activity is described in detail in Chap. 3

OD activities	CMR activities	CI activities
• Brainstorming • Scenario planning • Strategic planning • Innovation • Re-engineering • Resource allocation • Cost optimization • Downsizing	• Documentary research • Customer segmentation • Surveys • Focus group • Interviews • Big data analytics • Pricing	• Supply chain management • Due diligence • Benchmarking • Outsourcing • Mergers and acquisitions

client, and coming up with new ways to approach a problem is just good practice in management consulting. Table 1.1 lists 20 key consulting activities. Each of these will be described in more detail in Chap. 3 (Table 1.1).

2 Future of Big Data in Management Consulting

This chapter discusses the outlooks of management consulting, the interface with data science and the disruptive impact that new information technologies will have on the management consulting industry. The first part of this chapter presents key insights from the literature. The second part engages the reader into a scenario analysis that builds on these insights and starts imagining what the future of management consulting might look like.

2.1 General Outlooks in the Management Consulting Industry

As mentioned in the previous chapter in the section on the industry *Life Cycle*, probably the most important source of growth in the management consulting industry over the next few years will be an increased demand in emerging markets. This will be a solid source of sustaining growth for the overall industry, lead by China and India [9] who have so far outpaced expansion in the rest of the developing world. But recent trends also underscore more disruptive patterns of growth in the consultancy service portfolio, notably an increased demand for assisting in implementing the recommendations [51] and integrating IT based services [4]. These disruptions have started to materialize with the development of new internal capabilities (e.g. BCG's *Gamma*, McKinsey's *Solutions Practice*), and strategic alliances with specialists such as IT specialists, finance specialists, etc. In parallel and as a consequence, major firms in adjacent industries such as accounting firms Deloitte, PwC, KPMG, and IT firms IBM, HP, Accenture, are all expected to continue increasing the size of their consulting services [9].

Finally, it is expected that some challenges that started to surface in government client projects will expand to a broader client base: more clients will become more impatient for return on consulting investments and demand better evidences for success [4, 9]. Clients will increasingly expect that the recommendations delivered by the consultant be actionable right away, and because implementation is on the rise

J. D. Curuksu, *Data Driven*, Management for Professionals,
https://doi.org/10.1007/978-3-319-70229-2_2

and the fees paid for consulting services are not increasing proportionately, overall more value will be expected from a consultant's intervention.

In this chapter, let us focus on what is likely the largest disruptive force ever encountered in the management consulting industry. Indeed all the trends above toward enabling shorter, more repeatable processes that provide clearer ROI are at least partially lead by the revolution in the economics of information [2] and the joint development in computer-based data analytics [13, 52, 53]. But, as Harvard Professor Clayton Christensen recently noted [4], ever since the first management consulting company was created 125 years ago (Arthur D Little in 1886 [3]), the value proposition has always been grounded in deploying human capital and judgment-based solutions. Now this business model is being challenged by computer-based solutions. The future of computer-based data analytics in the management consulting industry is discussed here. Section 2.2 articulates key insights from the literature and Sect. 2.3 presents a scenario analysis that builds on these insights and starts imagining what the future might look like.

2.2 Future of *Big Data* in Management Consulting

The distinction between management consulting and more technical *IT* consulting services is becoming less and less evident, thanks to new business opportunities offered by the phenomena referred to as *big data* [2, 3, 13, 52–55] and *data science* [56–58]. Several drivers of innovation have been documented, some of which point to a potential disruption through modularization of a technology-assisted consultancy model [4, 49]. Whether sustaining or disruptive in nature, the threats and opportunities brought about by the democratization of information and big data technologies will impact the management consulting industry.

The literature points to at least five factors that may favor the integration of big data into management consultancy models, and at least three factors that may refrain its integration. Let us look at each of these factors.

2.2.1 Factors that Favor the Integration of Big Data in Management Consulting

1. *Big data technology dramatically increases the speed and accuracy of market research*

Today's businesses continuously create new data, from sales and marketing to customer relations, production, logistics, and more [2, 13, 52, 53, 59]. In addition, there's a huge amount of data generated by web sites, public databases, social media, etc. While traditional consulting methods may require weeks or months studying internal workflows, interviewing customers, or discussing with key personnel, one can now search social media, purchase history, and draw amazingly precise customer profiles in a matter of hours [53]. This may benefit both the consultants and the clients.

2. *Technology-assisted consultancy models enable modular and standardized fee-for-service offerings, providing clients with better control, faster delivery, and lower cost*

From the perspective of the clients, maybe the most attractive aspect of integrating big data into management consulting is the potential to solve a popular dilemma with consultants: that their traditional judgment-based interventions are difficult to control and cannot be reproduced because there is no definite standard [4]. In contrast, data science projects may be standardized, reproduced, and hence better controlled [56].

A management consultancy model assisted with advanced analytics technologies could become modular [4]. This model would be analogous to what software packages represent in the computer science industry: they offer a finite set of clearly defined and repeatable "sub-routines". Instead of paying for integrated solutions, including features that clients might not want, clients could supervise the problem-solving process, prevent consultants from reinventing the wheel with each successive assignment, and instead contract them for a specific analytics module, a specific link in the value chain [4]. For each intervention, the time for delivery and cost would represent only a fraction of expenditures associated with fully integrated projects. This would effectively transform the value proposition of management consultants from a fee-for-service model to a more flexible pay-for-output model.

This wave of commoditization would not benefit the management consultants. Even from the perspective of the clients, given the need to strike a balance between "reinventing the wheel" and "one size fits all", it remains to be seen whether and how much clients would benefit from such a disruptive innovation.

3. *Big data does not eliminate the need for traditional consultants. Even with big data, traditional business consultants are needed to ask the right questions*

Management consultants are used to drive their hypotheses using business models, acumen and inductive reasoning. Big data now offers the possibility to drive hypotheses and insights using real-time deductive reasoning [52], thanks to predictive analytics algorithms that may exploit patterns in big data within a few seconds. Combining these methods brings undeniable value to corporate organizations, but in many contexts nothing may fully replace the deep knowledge of business processes, markets and customer behaviors that consultants develop over time.

Predictive analytics may be used to identify risks and opportunities such as economic forecasts, cross-sell/up-sell targets and credit scoring. But the type of intuition that consultants develop to ask questions, pose hypotheses and drive executive decisions is still the realm of science fiction, not existing computer programs [60, 61]. Hence, the arrival of data scientists and big data analytics does not eliminate the need for traditional business professionals.

Table 2.1 Examples of data analytic providers offering business consulting services based on software capabilities

Type of Data	Example of Providers
General Consulting	Narrative Science, OpenIDEO, BeyondCore, Rokk3rlabs
General Marketing	Bluenose, Markertforce, Salesforce, Experian, Marketo, Genesys, Medallia
HR Management	Zapoint, VoloMetrix, Sociometric, Cornerstone, Salesforce
Frauds	Feedzai, Fico, Datameer, Lavastorm,
Surveillance	ADT, Frontpoint, Lifeshield, Monitronics
Driving	Zendrive, FlightCar, Progressive's PAYD, Metromile
Fitness	Discovery, Oscar, FitBit, Jawbone, Sleepcycle, Mealsnap
Health	Watson, Ginger.io, Sentrian, Aviva, AllLife, Kaiser Permanente, Flatiron
Emotions	Motista, Luminoso, Lexalytics, Xox, Watson

4. *New entrants embrace the new technology because it reduces brand-barrier to growth*

In contrast to the largest *generalist* management consulting firms, smaller firms (so-called boutiques) and new entrants must specialize in niche markets. But with *standardized* analytics softwares joining the toolset of management consultants, competition based on brand reputation is becoming less pervasive [4]. Factors such as product portfolio, technical capabilities, speed of delivery and convenience are becoming more relevant success factors.

5. *New data analytics technologies are already leveraged in many industries*

The big data innovation is already underway in many industries [13, 62–66], so management consultants will have to tag along. Potential clients process big data either in-house or through outsourced analytics providers, which sometimes works as a decent substitute to management consulting [66, 67]. For example, big data softwares have been developed to increase transparency between marketing performance and ROI [68].

Start-ups and subsidiaries are emerging with the sole mission of assisting corporate organizations leverage big data to optimize their businesses [53, 69]. This represents a threat to the role of *Analyst* in management consulting. Examples include assistance with structured data (e.g. how long a target market goes jogging every week?) and assistance with unstructured data (e.g. how much a product induces positive emotions?). Table 2.1 samples emerging organizations that recently met success with a computer-based data analytics business model to assist their customers with gathering the type of insights that used to be delivered by management consultants.

Large IT companies (eg. IBM, Accenture) are aspiring to become total service providers [57, 58]. This represents a threat to management consulting companies that do not internalize the new technologies. IBM is increasing investments in its "Global Business Services" and IBM Watson Cognitive System (launching "Watson-Health" in 2015), HP developed its "Business Service Management", and Accenture developed its own "Business Services" too. According to S&P Capital, management consulting in the IT industry will grow at an average CAGR of 7% [58].

The mirror phenomenon is taking place in management consulting firms: they are revisiting their service portfolio to assist clients with software development. McKinsey developed its "Solutions", Booz Allen Hamilton its "NextGen Analytics" and BCG is rapidly expanding "BCG Gamma". This is a potential threat to management consulting firms that do not internalize the new capabilities.

2.2.2 Factors that Refrain the Transition

1. *In some projects the nature of the data does not benefit from computer capabilities*

Management consultants often get essential insights based on just a few interviews inside/outside the organization, by looking at easy-to-digest financial records, etc. In these frequent cases, the data is not large and "big data" does not apply in the first place [60, 61, 70].

2. *Traditional management consulting revolves around executive decisions also when it pertains to data analysis. It prioritizes low volume, high quality, easy-to-digest data*

Management consultants drive executive decisions which are directional in nature. That includes all the steps before, during and after the data analytics activities. They are in charge of asking smart questions, navigating analytics tools, exploring data, interpreting data, building action plans, and increasingly helping implement these plans. As long as the consultant is involved in driving executive decisions, he/she will continue to follow the 80/20 rule and prioritize low volume, high quality, easy-to-digest data. Thus even when big data is available, the consultant might defer its use whenever a faster route to deliver insights is available.

In 10 years, the management consulting industry might have transitioned to a place where many clients redirect the consultant toward specific analytic tools. In this scenario the client would be doing a job currently held by the consultant. This would indicate that a disruption has taken place in the form of modularization and even commoditization. In contrast, if in 10 years data science has matured but using its software still requires a highly technical expertise that most clients cannot insource, then the currently emerging business models that blend core judgment based capabilities with technical capabilities such as McKinsey's Solutions and IBM's Global Business Services will have effectively disrupted the industry. Regardless, a disruption is underway…

3. *Internalizing highly technical tools that cannibalize traditional market research can create cultural dissonance*

Large global and generalist management consulting firms leverage their premier brand reputation (*cachet*) for contracting large clients on their most strategic, executive cases [71]. In contrast to smaller players and new entrants, larger organizations cannot fully embrace big data integration because it commoditizes their business, dilutes their focus, and weakens their brand.

2.3 So What: A Scenario Analysis

In this section, the future of technology-assisted management consulting is articulated using a scenario planning analysis. The scenario planning approach is concisely introduced in Chap. 3 (Sect. 3.1). More details can be found in Refs. [72, 73].

The Key Focal Issue
How will new data analytics and artificial intelligence technologies impact the business model of management consultants?

Driving Forces
In Table 2.2, some key factors and environmental forces are brainstormed and listed. This list does not aim to be exhaustive but rather to represent *pointers* toward forces that have potential to impact the key focal issue. This is standard procedure with the scenario planning framework of Garvin and Levesque [73].

Critical Uncertainties and Scenarios Creation
The driving forces listed in Table 2.2 may be condensed in two broad variables that I will refer to as *External Push* (from clients or consumers) and the future of *Partnerships*:

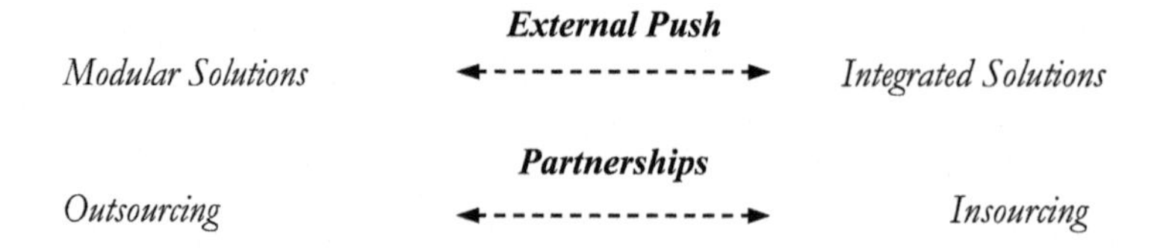

A 2 × 2 matrix may now be constructed with four possible scenarios:

Table 2.2 Key factors and environmental forces (*un-ranked*). This list of 13 forces results from simple brainstorming and is thus non-exhaustive, but the goal of scenario planning at this stage is not to be exhaustive nor precise, it is to list a *broad set of pointers toward forces* that may potentially impact the key focal issue

Driving Force	Level of Certainty
1. Client and end-customer push	Uncertain
2. Improvement in analytics software capabilities	Predetermined
3. Level of technical expertise required	Uncertain
4. Evolution of big data-enabled businesses	Uncertain
5. Modularization of consulting interventions	Uncertain
6. Budget reductions	Uncertain
7. Skepticism for traditional judgment-based analytics	Predetermined
8. Skepticism for commoditization and automation	Predetermined
9. Profitability of the McKinsey *Solutions* platform	Uncertain
10. Future of the IBM's *Watson* AI platform	Uncertain
11. Future of niche providers such as *Kaiser Permanente*	Uncertain
12. Future of portable sensor devices	Uncertain
13. Regulations (confidentiality, anti-discrimination...)	Uncertain

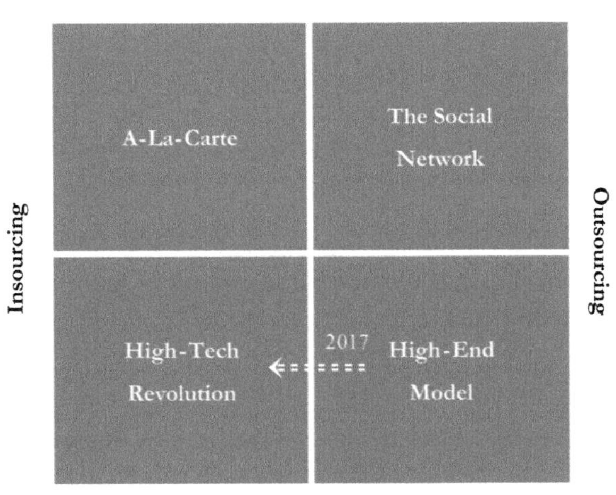

Narratives and Implications

For each of the four scenarios above, let us weave in defining characteristics and implications based on the insights gathered in Sect. 2.2 and the driving forces listed in Table 2.2.

"A La Carte"

A hypothetical future with Modular Solutions

1. Predictive technology and automation have increased to an extent that most executive decisions are now derived from big data
2. Big data analytics niche providers that flourished between 2010–2020 reached mainstream market, displacing generalist incumbents such as IBM and HP from their leading position
3. The type of technology-assisted management consulting that emerged during the big data revolution have cannibalized traditional business models that used to cover all aspect of a client's problem

A hypothetical future with Insourcing

4. Clients internalized the expertise needed for navigating the big data analytics arena and tend to contract consultants based on their menu of IT and analytics capabilities
5. Consultants struggle to preserve their identity of high-end strategy advisors in the marketplace, having become business data scientists, but the potential of big data insights provide new avenues for profit

"The Social Network"

A hypothetical future with Modular Solutions

1. Predictive technology and automation have increased to an extent that most executive decisions are now derived from big data
2. Big data analytics niche providers that flourished between 2010–2020 reached mainstream market, displacing generalist incumbents such as IBM and HP from their leading position
3. The type of technology-assisted management consulting that emerged during the big data revolution have cannibalized traditional business models that used to cover all aspect of a client's problem

A hypothetical future with Outsourcing

6. Consultants struggle to preserve their identity of high-end strategy advisors in the marketplace. They tend to be seen as middlemen between corporations and IT service providers. Clients still value their social and network skills but not their problem solving skills
7. Clients who internalized the expertise needed for navigating the big data analytics arena are integrating forward, which threatens the entire management consulting industry. These clients use the budget once allocated to management consultancy for specific IT analytics projects that, coupled with their own strategic insights, enable to solve problems efficiently and at lower cost

"High Tech Revolution"

A hypothetical future with Integrated Solutions

8. Clients have not internalized the expertise needed to navigate the big data analytics arena
9. Clients continue to rely on consultants for executive decisions which are directional in nature, based on easy-to-digest data analysis
10. Despite new possibilities offered by big data, clients are restrained by governmental bodies that act in the best interests of consumers and fight against coercion-like tactics. The growth of business models that leverage individual behaviors and portable sensors is limited

A hypothetical future with Insourcing

11. The new type of technology-assisted management consulting activities that emerged during the big data revolution have partially cannibalized traditional activities, such as surveys and focus groups, supply chain management and even documentary research! But the potential of big data insights provides new avenues for profit

"High End Model" …or forever executive

A hypothetical future with Integrated Solutions

8. Clients have not internalized the expertise needed to navigate the big data analytics arena
9. Clients continue to rely on consultants for executive decisions which are directional in nature, based on easy-to-digest data analysis
10. Despite new possibilities offered by big data, clients are restrained by governmental bodies that act in the best interests of consumers and fight against coercion-like tactics. The growth of business models that leverage individual behaviors and portable sensors is limited

A hypothetical future with Outsourcing

12. The big data analytics marketplace has become extra-ordinarily fragmented during the big data revolution that took place in 2010–2020. The need to help customers navigate through the big data analytics arena gave a boost to the management consultants' portfolio
13. More than ever, management consultants focus on their core competencies. They leverage traditional business models based on integrated solutions that cover all aspects of a client's problem, including executive decisions, and rely on plain outsourcing for most data analyses

14. The big data revolution has cannibalized several traditional activities, such as surveys and focus groups, supply chain management and even documentary research
15. The industry's future is uncertain. The *Analyst* role has become obsolete, and the number of *Consultants* has decreased due to outsourcing of many activities. Academics become direct competitors to management consultants because they combine flexible profession with technical expertise

Early Warning Signals

In order for managers to recognize events that may signal the emergence of one scenario versus another, let us propose a series of "early warning signals". For example, the success/failure of the McKinsey Solutions platform will be a strong predictor of whether premier consulting firms will have insourced or outsourced the disruptive technology by 2020.

- Success/failure of the McKinsey Solutions platform
- Significant market share gained by niche players against generalist incumbents in the big data analytics market place
- Publicly acknowledged shift in strategic priorities of large corporate clients on how they plan to leverage big data
- New governmental bodies or regulations that aim at changing the way data is used or disseminated in order to protect consumers against discrimination

Epilogue

Each of the four narratives derived in this chapter, using a literature review (Sect. 2.2) combined with a scenario planning analysis (current section), represents an alternative possible outcome for the management consulting industry. But how new information technologies will *ultimately* impact the business model of management consultants is likely not any of these four alternatives but rather a combination of some of them, and will of course include additional unforeseen factors.

Nevertheless, one may reasonably expect that the future will borrow some elements of these scenarios and even be biased toward one of these four scenarios.

Hopefully, this brief analysis helped the reader frame the scope of some future developments to come. Managers in management consulting and client companies are invited to develop action plans based on what they believe is the most promising/likely in these scenarios, monitor events as they unfold – watching for early warning signals – and prepare to change course if necessary. Because this is all what a scenario planning analysis claims to offer, an aid to thinking about reality, without arrogance, as for consultants.

Toolbox of Consulting Methods 3

This chapter introduces methodological aspects of some consulting activities. The reader is advised to complement insights gathered from this sample of *methods* with the set of theoretical, overall *frameworks* introduced in Chap. 5. Indeed, if current chapter is about what consultants do, Chap. 5 is about how consultants decide which methods to use for a given case, team, client and circumstances.

As mentioned earlier, this chapter will not delve into the different chronological phases of an assignment because this has more to do with developing a client-consultant relationship than developing operational capabilities. The client-consultant relationship is the focus of Chap. 4, and for this reason the chronological phases of an assignment will be covered in Chap. 4 (Sects. 4.4, 4.5 and 4.6).

Finally, the methods presented in this chapter do not represent an exhaustive set. Consultants may find here a set of pointers, a backbone if you will, for action planning. Consulting activities may be organized in three categories – organizational development (Sect. 3.1), consumer market research (Sect. 3.2) and competitive intelligence (Sect. 3.3). The first category regards issues internal to the client organization, the second issues with external stakeholders and the third issues with market players (who may represent competitors, partners, or acquisition targets). Defining these three categories is useful to organize the discussion, but of course the merit is purely pedagogic since any activity may be adapted to fit the needs of particular circumstances, and coming up with new ways to approach a problem is just good practice in management consulting. Many more methods are available, and each method will find contexts in which it may be used in ways not considered here. But for the purpose of an introduction, the reader will find in this chapter a set of 20 action-oriented recipes. These can be accessed interactively on *econsultingdata.com*.

J. D. Curuksu, *Data Driven*, Management for Professionals,
https://doi.org/10.1007/978-3-319-70229-2_3

3.1 Organizational Development

3.1.1 Strategic Planning

One definition of strategic planning is the determination of an organization's objectives together with a comprehensive set of actions and resources that are required to meet these objectives. Strategic planning is a process that assists executive managers in their resource allocation and decision-making [74].

Methodology #1

1. Articulate the client's mission, vision and values
2. Identify and evaluate the potential of the client's currently targeted customer segments
3. Explore threats and opportunities from a customer perspective
4. Benchmark competitors
5. Define the client's strengths and weaknesses relative to its competitors for each segment
6. Develop a business model that will offer competitive differentiation or cost advantage
7. Articulate milestones and objectives that relate the proposed model to the client's mission, vision and values
8. Describe a set of programs, organizational structures, policies, information sharing systems, control systems, training systems, that will implement the proposed model
9. Identify, evaluate and present alternative models (see *scenario planning* in Sect. 3.1.4)

3.1.2 Innovation

Developing an innovative product, service or business model requires developing a strategy and juxtaposing it with the client's capabilities. Innovation can come from internal innovation *task forces*, where consultants may assist with initiating or improving internal processes that foster open circulation of innovation ideas. It can also be brought in from external players (outside-in innovation), through strategic alliance or (more rarely [47]) mergers and acquisitions. In this case consultants are more likely to engage in market research and due diligence activities.

Discovering innovation opportunities may be attempted using any number of primary and secondary market research activities, and watching for *warning signals*. Researchers who specialize on growth and innovation [18, 19, 46, 75, 76] recommend to go beyond what customers "say" and directly observe what they "do". Observing their consumer decision journey can uncover compensating behaviors, workarounds, and pinpoint inadequacy of existing solutions. These disparities are the hallmark of innovation opportunities.

The concepts of *jobs-to-be-done* [18, 19, 46, 47, 75], *blue ocean* [77] and *disruptive innovation* [18] for example, provide outstanding frameworks to any effort aimed at discovering opportunities that arise from non-consumption. In targeting *non-consumers* who face a barrier inhibiting their ability to get a job done, competitive responses from existing players is minimized because the innovation responds to the need of a new market, a market that does not fit the definition of any pre-existing company's target market.

Disruptive innovations create new markets, but also transform existing ones though simplicity, convenience, affordability or accessibility, addressing the need of *over-served consumers* [18, 19, 46, 47, 75]. Innovating does not necessarily require a new product: an innovative business model (how a company creates, captures and delivers value [76]) may disrupt markets as severely as any new technology. It suffices to look at companies such as Uber, Netflix, Google and Amazon to see how business model innovation with already existing technologies can re-shape the dynamics of entire industries.

Methodology #2

1. Coordinate brainstorming sessions between executives and middle managers for innovative ideas. Alternatively, the consultant may gather proposals directly from them. But managers should be made comfortable discussing ideas without the usual organization self-imposed financial and operational restraints, because they are the closest to day-to-day value adding activities
2. Gather potential outside-in innovation ideas from other players in the industry and adjacent industries, using due diligence activities
3. Determine for each suggestion whether it represents an incremental or a disruptive innovation
4. For incremental innovations, decide upon projects with most potential based on highest return, lowest risk and best fit/synergy with current core competencies
5. For disruptive innovations, analyze the costs and benefits under alternative assumptions (see scenario planning in Sect. 3.1.4), get top-level commitment by emphasizing the threat of the disruption, and create an autonomous subsidiary organization that will manage the new business initiative. It is essential that the subsidiary organization frame the innovation as a top growth opportunity (not a threat)
6. Develop internal information systems and teamwork incentives to enable the client in the future to capture insights from executives, middle managers and employees
7. Develop external processes to facilitate outside-in innovation in the future, such as initiatives aimed at building a reputation as an innovative partner and alliances with innovative partners

3.1.3 Re-engineering

Re-engineering (a.k.a. *process innovation*) was pioneered by MIT Professor Michael Hammer in the early 1990s [78], and gained traction thanks to the support of established management figures such as Peter Drucker [79]. In parallel with popular approaches referred to as *Lean* and *6-Sigma* [80], the goal is to address customer needs more effectively and/or at lower cost, and enable non-value adding functions to effectively become obsolete. In contrast to its cousins Lean and 6-Sigma, re-engineering emphasis is on developing *cross-functional* teams and better capabilities for *data dissemination* [81]. It generally involves analysis followed by radical redesign of core business processes [82] to achieve dramatic improvement in productivity[1].

Methodology #3

1. Articulate a focused customer need and how the client's organization addresses it
2. Rethink basic client's organizational and people issues
3. Redesign a workflow using information technology
4. Reorganize the workflow into cross-functional "agile" teams with end-to-end responsibilities

3.1.4 Scenario Planning

Scenario planning [72, 73] helps organizations handle unexpected circumstances as they unfold while focusing effort on the circumstances that are most likely to take place in the future. The consultants and managers brainstorm on "what-if" hypothetical environments and develop appropriate courses of action for each alternative. For a detailed example, see Chap. 2 (Sect. 2.3) where the method was applied to delineate how new information technologies are eventually going to disrupt the management consulting industry.

Methodology #4

1. Define key questions and challenges in order to articulate a key focal issue
2. Research driving forces that impact the key focal issue, and distinguish forces that are predetermined versus forces that are uncertain
3. Rank uncertain driving forces starting with the ones that are most uncertain and influential.
4. Derive two critical uncertainties, each of which may represent one or several driving forces
5. Create a 2×2 matrix from the two critical uncertainties and develop a narrative for each quadrant. For each scenario, start with a catchy headline and then frame a consistent plot line by weaving characteristics and driving forces into stories

[1] The *radical* level of change implied in re-engineering was elegantly contrasted with the more *incremental* level of change implied in TQM (total quality management) in Ref. [82].

that match the headline. To build compelling stories, step through the fictional development that leads to the headline using newspaper styles, popular characters/organizations and standard fiction plot lines
6. Generate and prioritize implications for each scenario
7. Develop action plans based on most promising and likely scenarios
8. Prepare the organization to monitor events as they unfold and watch out for early warning signals
9. Prepare the organization to change course if necessary

3.1.5 Brainstorming

Collecting information around a theme from just a few persons can be surprisingly efficient when coordinated using a structured approach. Brainstorming may take very different forms depending on the type of issue (e.g. open innovation vs. cost optimization) and the type of participants (e.g employees vs. customers). A *focus group* [83] for example is a common brainstorming activity applied to market research, and described in Sect. 3.2.4. The step approach below represents a more "general" recipe for brainstorming.

Methodology #5
1. Gather a small group (e.g. 5–10) of people with different expertises and leadership roles, including one person who will just take notes (*idea recorder)*
2. Clearly present the objective of the meeting
3. Present a set of initial ideas around the topic of interest
4. Coordinate a round table where each participant provides feedbacks on what has been said so far and submits his/her own ideas. In particular, make sure that only one person speaks at a time, eventually requesting participants to put their hands up to request to speak in the future
5. Encourage and promote creativity by regularly suggesting specific out-of-the-box thinking, e.g. one may ask how would a problem be dealt with in another industry, by Clay Christensen, Steve Jobs, Warren Buffet, a blue collar, a foreign citizen, etc.
6. Create a *car park* that stores ideas that don't quite fit with what is being currently discussed and revisit these later
7. Finish with a discussion based on mutual feedbacks that aims at building consensus and efficient decision-making

3.1.6 Resource Allocation

Budgeting is key to boost competitiveness as it aims to optimize the alignment of resource allocation with strategic goals. As for re-engineering, making effective resource allocation decisions requires a deep analysis of which activities should be eliminated, which activities should be performed, and how to perform these activities. It requires a quantitative costs/benefits analysis of alternative options such as automation, streamlining, standardization, outsourcing and offshoring. As for many financial estimation techniques, the ROI of certain activities such as R&D may be difficult to estimate and might sometimes make sense only if the client is receptive to long-term returns [84].

Methodology #6

1. Identify clear strategic plans and cost targets
2. List facts and figures for the different activities and business units, including all expenditures and how they relate to revenue streams
3. Define an *ideal* state and *essential* activities
4. Optimize or re-engineer the future state and set of activities
5. Ensure client acceptance and reset budget accordingly

3.1.7 Cost Optimization

To reduce costs and maintain growth, inflection points need to be found where products/services can still fully meet customer demand but at lower cost [85]. Cost optimization solutions often consist of simplifying the organization structures (e.g. operating models, marketing channels) and innovating new technologies, business models or IT systems [86].

Methodology #7

1. Identify activities that are potential tradeoffs between necessary competitive edge (e.g. product customization) and unnecessary complexity (e.g. middle management)
2. Brainstorm and determine alternative solutions for these activities (e.g. streamlining, outsourcing, restructuring, vertical integration) that may reduce cost and yet strengthen the core (e.g. increased focus, improved brand exposure, accelerated production)
3. Perform quantitative cost/benefit analyses to compare these alternative models
4. Develop and present action plans for the most beneficial alternatives

3.1.8 Downsizing

As ultimate mean to reduce costs, the client might consider downsizing. The consultant then needs to revolve its entire analysis on the following question: *What non-essential assets can be eliminated in order to minimize negative impacts on the remaining organization*?

The many drawbacks of downsizing [85] include poor public relation, damaged employee morale, and wasted opportunity for capitalization if economic circumstances improve in the future (i.e. it is more expensive to hire and train a new employee that to incentivize an existing one). Hence everyone, customers, employees and executives, have something to loose in downsizing. As a consequence, the consultant faces a dilemma that lies between cost reduction and effective growth strategy. If the consultant proposes an efficient downsizing strategy that results in side effects more damaging that the problems it manages to cure, the client's bottom line will suffer and the client-consultant relationship deteriorate.

Methodology #8

1. Identify nonessential company resources as potential candidates for downsizing
2. Evaluate the overall impact of downsizing each candidate on the remaining organization, including both financial and psychological impacts. Particular attention needs be brought on the company's reputation, lower employee productivity due to disorder or talent loss, and future opportunity costs in case that economic circumstance permits the organization to grow back to its current size
3. Ensure client acceptance
4. Develop a *program*, which may include: building a committee to support the job cut and relate it to the best interests of the company and its shareholders, training managers to conduct layoffs, assisting former employees in their job searches

3.2 Consumer Market Research

3.2.1 Documentary Research

Consultants extract information from literature, reports and internet resources at every stage of an assignment. They use this type of *secondary research* [87] to get up-to-speed before an initial client meeting and before/during all types of activities (OD, CMR, CI) because whichever problem is addressed, it has probably already been addressed before. It is important for consultants to follow the *80/20 rule* [88] and strike a balance between big picture (not enough) and very detailed (too much) information.

Methodology #9

1. Articulate the theme being researched
2. Define a list of keywords, including synonyms and alternative terms
3. Establish an appropriate scope for the research: type of publication (peer reviewed articles, activity reports, industry reports, newspapers, books, blogs), date of publication (<1 month, <1 year, <10 years), depth of publication (summary, full text, individual sections)
4. Interrogate relevant databases using keywords and scope defined above
5. Select a set of documents by exploring table of contents, summary, charts
6. Exploit the selected documents: an efficient method is to focus on the questions and theme being researched at all time while reading, highlight sections that relate to these particular questions and particular theme, and at the end of each document synthetize takeaways that will enable easy access in the future

3.2.2 Customer Segmentation

Segmenting a market is important for an organization's ability to focus and develop uniquely appealing offerings tailored to the most profitable customers, effectively outperforming competition in a particular target market. The competitive advantage does not necessarily have to take the form of enhanced product features: it may instead involve savvy marketing campaigns or dynamic pricing strategies. Customer segmentation helps companies in their resource allocation process and value adding prioritization scheme. It may be carried out based on any number of customers' demographic or psychographic attributes (age, income, ethnicity, education, gender), but an increasing number of researchers have expressed skepticism with this approach [18, 19, 46, 75] and instead propose to segment customers based on the circumstances under which customers proceed with a purchase, the reason why they buy the client's product (a.k.a. *jobs-to-be-done* [18, 19, 46, 47, 75]).

Methodology #10

1. Identify measurable market segments that are meaningful to the client's value proposition
2. Carry out (or use results from) customer surveys, focus group and 1-on-1 interviews
3. Determine the profit potential of each segment by analyzing the cost and benefit of serving each segment
4. Prioritize segments according to profit potential and client capabilities/ objectives
5. Invest client's resources to match offerings (product differentiation, pricing, marketing, distribution) with the characteristics of the selected segments
6. Prepare organization to adjust its offerings to future changes in market conditions

3.2.3 Surveys

Surveys are frequently used to assess thoughts, behaviors, opinions, needs and jobs-to-be-done of various stakeholders inside and outside the client's organization. Contemporary survey methodologies take the form of questionnaires that lend themselves to computer-based statistical analysis. For example close-ended questions, where the respondent picks up an answer from a limited number of options, may be analyzed using simple descriptive statistics (see Chap. 6 for details). But open-ended questions too may now be automatically processed by a computer using state-of-the-art analytics technologies referred to as *natural language processing* [89] (see Sect. 7.4.3 for details), or simply be coded into a response scale that will subsequently be treated quantitatively. A survey generally contains both close-ended and open-ended questions. The first type is easier to record and score but diminishes expressivity and spontaneity [87]. Open-ended questions allow responders greater flexibility, but are difficult to record and score.

A survey contains many features that may potentially be improved, thus careful design and interpretation are essentials. For example, to generalize the findings from a sample to a population, it is necessary to avoid the so-called *selection bias* (i.e. over-representation of a given gender, race, education level, etc). Random and stratified sampling are often used to minimize selection bias, where *stratified sampling* consists in dividing the population into sub-populations and drawing a random sample from each of these "strata" [90].

Another common challenge of statistical surveys is the presence of *spurious relationship*, also named *correlation-causation fallacy* [90], where an observed correlation between two variables may be explained by a third (often unknown) variable. The consultant might be tempted to develop a strategy or action plan based on the assumption of causality between two variables x and y that he/she found to be correlated. But if variable x is not "causing" variable y, if both variables x and y in fact depend on some other variable entirely (call it z), the intensity of the correlation between x and y might vary depending on the value of z. For example, assume x and y are two consultants' moods, and z is their manager's mood. For some range of the manager's mood (e.g. happy), the consultants' moods may strongly correlate as they experience joy and stress in concert. But for some other range (e.g. unhappy), their mood may become completely uncorrelated: they are now more influenced by factors such as which project they will *separately* be staffed on next than by this current *common* project. Mathematically, this type of conditional relationship is quantified using *partial correlations* [90] (see Chap. 6 for details).

A less well-documented challenge of the survey methodology is the unreliability of generalizing past and present to predict the future [91]. It is common in newspapers to read about "attributes" of a population inferred from a "rigorous" survey, when in fact these attributes are *dynamic* variables and thus their past trajectory may hardly repeat in the future[2]. In a so-called *cross-sectional* study, the survey is carried-out at one particular time; in *successive independent* studies, a survey is carried out at multiple time points, attempting to draw samples that are equally representative of the target population; in *longitudinal* studies, a survey is also carried out at multiple time points but on the *same* sample. Longitudinal studies are the most rigorous [91]. But even a perfectly orchestrated longitudinal survey will not fully predict the future.

Predicting a company's bottom line in a year is akin to forecasting the weather in a month. The past trajectory is indicative of its future, but any forecast must come with an associated probability [92]. The consultant should always identify a level of confidence.

Methodology #11

1. Decide what kind of information should be collected, and between *1-time* vs. *longitudinal* studies

[2] Even with last generation big data analytics software. For examples, read the WSJ article titled "*Economic Forecasts, Without the Lag Time*" by Josh Zumbrun published on August 25, 2015.

2. Construct a draft of questions. Questions should be simple and direct. Start with questions that catch the respondent attention and move demographic questions to the end
3. Pre-test the questionnaire with a pilot
4. Edit the final questionnaire and specify instructions for respondents
5. Decide upon classification criteria (e.g. needs, jobs-to-be-done, age, gender, location, education, profession, income) and with these criteria select a representative sample of a target population using the sampling techniques described above (e.g. *stratified sampling*)
6. Develop an action plan to increase response rate and reliability. For example, introduce the survey with friendly instructive comments (e.g. length, goal, interviewer's organization name), send a pre-invitation to assess cooperation beforehand, provide incentives (gift, lottery), test-retest to assess consistency of the respondent's responses, and avoid the *interviewer effects* described in Sect. 3.2.5
7. Complement insights from verbal surveys with *direct observation* to determine with better accuracy the behaviors that stakeholders really engage in, the attitudes they truly uphold

3.2.4 Focus Group

Focus group is a type of brainstorming activity that fosters diversity and creativity while revolving (*focusing*) on a predefined theme [83]. It is traditionally used as a market research tool to understand how customers will react to a new offering [87], but it may also be used in any circumstance where diversity and creativity are high priorities. In short, a focus group consists in selecting a diverse set of participants and implementing a process that prevents conformity and oblige each participant to develop and express his/her own independent opinions.

Methodology #12

1. Gather a small group [19, 46, 76–79] of people with diverse background, such as different education/income levels or ethnicity
2. Present the entire set of questions/assertions around the theme that will be addressed
3. Coordinate a first round table based on a simple rating system, such that each respondent develops an opinion and commits to it, but does not express it and thus does not influence others (this is the key step, adapted from Ref. [93])
4. Coordinate a second round table with a particular focus on questions/assertions that created the most diverse answers, and invite each participant to express the rationale behind its rating in the first round
5. Now that diversity has been fostered, coordinate a discussion based on mutual feedbacks that aims at building consensus and efficient decision-making

3.2.5 Interviews

Interviews are often used instead of a survey because they allow the interviewer to increase response rate and reliability [87]. They can be the only reasonable format in some circumstances (e.g. meeting with a CEO). Most importantly, interviews may include *direct observation* of certain behaviors and attitudes that can lead to completely new insights [18, 46].

Interviews cannot reach the scope of surveys that draw over multiple questions over hundreds of respondents of course. But even in these cases, interviews may be necessary to complement the regular survey because some of the questions might require face-to-face interaction. For example, in biomarker studies over medical patient cohorts, some questions often require complex diagnostics that can only be carried out by an expert.

A common challenge with interviews is the *interviewer effect*. For example in a face-to-face survey, participants may attempt to project a positive self-image in an effort to conform to the norms they attribute to the interviewer asking questions [94].

Another common challenge with interviews is the *mode effect*. Whether the meeting takes place under formal hierarchical corporate structure, in-person at home, by street intercept, by telephone, or by text online, may change how the respondents answer [95].

As for surveys, the questions should be simple and direct, and contain both open-ended questions which allow responders greater flexibility but are difficult to record and score, and close-ended questions which are easier to score even though they diminish expressivity and spontaneity.

Methodology #13

1. Articulate the objective of the interview and what kind of information should be collected
2. Prepare a discussion canvas to structure the meeting [93]. The opening could include the objective of the meeting, the body could be categorized by theme, each of which would direct to a specific set of questions, and the closing could solicit documents/contact information and indicate next steps, i.e. when the interviewee will hear back from the interviewer. In contrast to surveys, starting with simple demographics questions is useful to boost the respondent confidence
3. Start carrying out the interview by breaking the ice, introducing the objectives and proposing an agenda for the meeting. Proposing an agenda makes the consultant appear rigorous but also sensitive: the interviewee feels more comfortable when a clear beginning, objectives and milestones, have all been prepared in advance
4. During the interview, focus efforts on active listening and direct observations of behaviors, attitudes, purchase journey, etc. Take note of key ideas and verbatim quotes
5. Write a detailed report of the interview once the meeting has come to an end

3.2.6 Big Data Analytics

Big data is a popular buzzword today [96, 97] that reflects the explosion of large scale data readily available through internet in virtually every domain of application (the twenty-first century is *the new age of information* [98]). *Analytics* refers to the statistical and mathematical techniques used to analyze all these data [52]. The newborn science, *data science*, is becoming increasingly accessible to everyone without a need for expert-skills [98]. Big data analytics is changing the best practices of measuring and managing critical functions within a continuously expanding net of business areas [53, 99]. The potential opportunities/challenges in big data are commonly referred to under the 4-Vs [65]:

- Volume: Larger amount of accessible data
- Velocity: Fast/ultra-fast analysis, opening the door to real-time processing
- Variety: Any type of data, from structured spreadsheets to unstructured texts, sounds, speeches and videos
- Variability: Combination of different data formats into predictive models and artificially intelligent systems

Methodology #14

1. Articulate some objectives, identifying critical functions and performance measures that will be used to evaluate success
2. Determine appropriate data sources, analytic/AI solutions and resources
3. Ensure client acceptance
4. Form an internal or external team of technically skilled data scientists
5. Develop policies related to processes and data confidentiality
6. Facilitate the collection of data between relevant parties and the data science team
7. Develop a business strategy based on insights gathered from predictive models and artificially intelligent systems
8. Develop a framework to enable the client to continue testing and learning with big data

3.2.7 Pricing

Pricing strategies should factor in the client's objectives, its cost structure, the competitive landscape and the market's elasticity of demand (a.k.a. *economic value* [100]). Pricing optimization techniques are closely connected to product design because different customer segments generally look for different products, which in turn correspond to different *price points*. Under a predetermined set of assumptions, complex mathematical models may be developed that adjust prices dynamically to factors such as inventories, seasonal demand, market players' innovations (e.g. software upgrades). Given the *precise* nature of these models' outputs yet the often volatile and *inaccurate* nature of their inputs (e.g. assumptions, survey results), it may always be better to complement pricing strategies with the scenario planning approach introduced in Sect. 3.1.4. The outcome then takes the form of a continuous price curve on which the client may choose to position itself based on its own beliefs and favorite assumptions.

Methodology #15

1. Articulate the client's objectives (short term profit, market share, or other strategies)
2. Select the preferred optimization model and determine required inputs/outputs
3. Collect data necessary to run the model. For example: cost structure, competitors cost structures, economy of scale, seasonal balance, competitive response, new entrants, new technologies, impact of marketing campaigns, customer surveys and interviews
4. Load, run, and revise the model
5. Implement pilots to gain more confidence in the model's predictions
6. Describe processes that will permit to implement the proposed pricing strategy
7. Develop a framework to enable the client to use future data to upgrade the pricing model

3.3 Competitive Intelligence

3.3.1 Supply Chain Management

This activity consists in synchronizing and optimizing the consecutive phases that a product or service passes through from procurement to delivery [86]. An organization's operations may be divided into two parts: the *value chain* and the *supply chain*; while the value chain defines the chain of successive activities that add value [100], the supply chain refers to the logistics that enable these activities to take place. In other words, supply chain management is what gets the right product to be delivered at the right time, for the right cost, at each phase of the value chain. Effective communication and synchronized information transfer between all parties involved are thus essential to supply chain management [86].

Methodology #16

1. Develop a team building-like exercise between the client and different partners to build trust and open communication
2. Identify what key data are needed from each partner for its optimal integration into the client's value chain. For example: delivery dates, production schedules, demand forecasts, capacity utilization, inventory levels
3. Implement automatic data exchanges IT systems and incentive processes that will ensure the optimal transfer of key data from each partner identified in stage 2
4. Supply chain management may also involve more radical design initiatives, such as the re-engineering method [78–81] described in Sect. 3.1.3

3.3.2 Due Diligence

Due diligence refers to an ensemble of investigative techniques used to define a given organization's performance. It is often used prior to a merger or acquisition where the consultant seek to recommend one or just a few interesting targets for the integration, based on the client's objectives. One essential contribution from the consultant is often in developing a comprehensive set of performance measures that can be quantified and appraised [86].

Methodology #17

1. Articulate a clear vision and strategy for the client and targets' businesses.
2. Identify performance categories (e.g. financial, operational, innovation) and objectives
3. Develop measures of performance that take into account short-term and long-term results
4. Ensure client's acceptance of the measures
5. Create sound tracking and communication systems
6. Collect and analyze performance data

3.3.3 Benchmarking

The goal of benchmarking is to innovate by incorporating *best practices*. It is not just an imitation game. It consists in identifying, understanding and learning from alternative processes and practices that drive superior performance (in operations, sales, etc). This may be done internally by comparing similar activities within the client's organization, or externally by looking at competitors and in particular the *best-in-class* (i.e. best performing) companies [85].

Methodology #18

1. Identify key performance metrics
2. Choose companies or internal areas to benchmark
3. Collect data on performance and practices
4. Analyze data and identify opportunities for improvement
5. Adapt best practices to client's core competencies

3.3.4 Outsourcing

Outsourcing enables a company to focus on its core competencies more efficiently. The contracted third party company's business model generally leverages focus and scale, as it specializes in the outsourced activities. The outsourcing company in turn benefits from relatively low cost and state-of-the-art activities without having to internalize them [101].

Methodology #19

1. Determine which activities correspond to core competencies and which activities might be outsourced
2. Evaluate quantitatively the cost and benefit of outsourcing different activities
3. Assess the non-financial impact of outsourcing different activities, for example potentially lost synergies and dependence on the contracted outsourcer
4. Rank the best opportunities (and potential outsourcers if several options are available)
5. Ensure client acceptance
6. Develop a contract with guidelines and performance measures that fit client's expectation

3.3.5 Mergers and Acquisitions

A consulting project in mergers and acquisitions (M&A) may take very different paths depending on the nature of the M&A (e.g. consensual acquisition, hostile takeover, bona fide merger) and the phase at which the client calls upon the consultant (e.g. to identify potential targets versus to facilitate the integration).

Identifying targets will focus the consulting assignment on due diligence (Sect. 3.3.2) and strategic planning (Sect. 3.1.1) activities. Assisting with the integration in contrast will involve more operational activities such as supply chain management (Sect. 3.3.1), benchmarking (Sect. 3.3.3) and most OD (Sect. 3.1) activities. Regulatory frameworks can have dramatic impact on the initiation and well being of large M&A ventures as they may be deemed anti-competitive and thus require an understanding and support of governmental organizations (e.g. FCC in the US).

Most M&As fail. Researchers identified a number of biases (both in the screening and integration phases) that lead to these failures, including the *confirmation bias* where the consultant seeks confirming evidences but not disconfirming ones, an *overconfidence with assumptions* made, the *lack of contingency* plans ("*sunk cost fallacy*"), the underestimation of *cultural differences* and the underestimation of *time and resources* needed for the integration [102].

Methodology #20

1. Articulate the M&A objectives (e.g. penetration, diversification, synergies, tax benefits)
2. Review/select target company(−ies) using due diligence, including financial forecasts
3. Develop a plan for the integration, identify specific milestones
4. Form an integration team that will oversee the integration
5. Communicate clearly the strategic motivation behind the M&A to all employees
6. Develop incentive programs to encourage all employees to commit to only one culture
7. Get support from both organizations and their immediate contribution to the integration

The Client-Consultant Interaction

4

4.1 Nature of the Relationship

Before to discuss client expectations (Sect. 4.2) and interactions along the different phases of a project (Sects. 4.3, 4.4, 4.5, and 4.6), it may be useful first to clarify what is the nature of the relationship between a consultant and his/her client. Different types of relationships have been articulated in the literature. As you will see, the nature of the relationship depends on the context and most particularly on the client's *perception* of what represents success in a consultancy. The consultant should thus beware of these different perceptions and adapt to the situation.

4.1.1 The Big Picture: Theoretical Models

No less than a dozen of broadly accepted models of consultancy may be found in the literature, illustrating contrasted viewpoints on the profession. To first approximation, all these models may be clustered into four categories on the basis of how much expertise the client acknowledges in the consultancy. A sort of continuum will then emerge: it starts with "total" reliance on the consultant, ends with "total" skepticism, and passes through more balanced views.

1. *The classic expert model*

In this model the client transfers information to the consultant and the consultant ultimately hands over solutions. It assumes a consultant's capacity to solve the client's problem [103]. This is a consulting-centric view, thus quite unidirectional and ideological. Nevertheless, it underscores two core reasons that made the consulting industry become so successful, which were suggested more than 30 years ago by Peter Drucker [104] and have so far stood the test of time: external consulting can be impartial because its agents are independent, and can bring new ideas because its agents are exposed to different industries/companies.

J. D. Curuksu, *Data Driven*, Management for Professionals,
https://doi.org/10.1007/978-3-319-70229-2_4

Other names found in the literature for this model are the *purchase of expertise* [105], the *mental adventurer* and the *strategic navigator* [106]. From a management perspective this model arises from the *classical* organization theory: it is rooted in some highly normative essays that go back to Adam Smith and some early twentieth century's popular economic theorists [107]. These essays paved the way for the formulation of rules and how-to guides in the pursuit of "scientific" management, successful, and ultimately "perfect", consulting. But again, this theory is widely criticized for its ideological tone [103, 48, 106, 107].

2. *The doctor-patient model*

This model is more balanced in that the client-consultant interaction is a joint learning process [108]. While the expertise of the consultant is omnipresent, it refers to an expertise in management and processes rather than in the client's concrete problems. The popular analogy with a medical doctor comes from the focus on the diagnostics of the client's problem. It ensues that the consultant's capacity to solve the client's problem gets "constructed" by a joint effort between the client and the consultant. Both are equally resourceful for developing the solution. It underscores the importance of developing comfort and mutual trust in the interaction [7].

Other names found in the literature for this model are the *business doctor* [109], the *management physician* and the *system architect* [106]. From a management perspective this model arises from institutional economics paradigms [110] themselves connected with (and rooted in) evolutionary life science paradigms, most particularly system biology. In these paradigms, individuals strive to maximize their economic benefits in incomplete market situations, and institutions steer individual behaviors into certain directions in pursuit of specific goals [107]. Natural friction arises because individuals and systems (institutions) have different incentives, which may lead to potential opportunistic behaviors of either partner [107], and because a system is defined as much by its inner elements as by how these elements contrast with outside ones [111]. A fundamental postulate of system theory indeed is that changes in a system are "self-organized" and thus cannot be managed from the outside. This naturally creates friction between system (client) and non-system (consultant).

This model doesn't acknowledge the existence of an all-embracing consulting expertise, but does not contradict the existence of an expert system that specializes in processes, routines, strategies, and hence who possesses unique diagnostic capabilities. The "doctor" will develop treatment plans; the "patient" will choose what specific solutions fit its interests. This theory underscores the need for developing effective terms of cooperation, so that the benefit of the consultant intervention outweighs the natural friction of this intervention on the client system.

3. *The social helper model*

In this model the client-consultant interaction is also a joint learning process, but, simply put, the client is the expert. The consultant's value proposition lies in the confines of psychology and sociology frameworks. It might be best understood as *assisting* the client with solving a problem, therefore assuming no responsibility in this journey.

The consultant facilitates the problem-solving journey [103] within the organization by building on Drucker's first value: the consultant is an independent party thus it can more easily facilitate teamwork and exchange of opinions within the organization, foster motivation and acceptance. In addition, the consultant contributes to these discussions by building on Drucker's second value: the consultant benefits from exposure to different industries/companies that internal experts do not possess, thus he/she has a unique potential for sharing new perspectives.

Other names found in the literature for this model are the *reflective practitioner* [112], the *social learning model* [113], *process consultation* [108] and the *friendly co-pilot* [106]. From a management perspective this theory is rooted in organizational development and sociological sciences, and rationalizes the widespread dissatisfaction with the expert model of consultancy. It goes beyond a rational "Socrates" approach to solving a problem [114], in that it considers hidden cultural, political and emotional factors as intrinsic elements of the problem to be solved [115].

The core idea in this relationship is that the solutions to the client's problem will come from a social experiment, where it is the client who brings most of the expertise, while the consultant brings *methodologies* to engage experts and assist in the confrontation of ideas. If the expert model is the conservative view of management theory, then the social helper model is its antipode, the "open" view. But as one might expect, drawbacks include frequent bias entailed by political agenda in what comes out of the problem-solving journey, at the cost of pragmatism [7, 115]. This model starts questioning the legitimacy of the industry as a whole, and opens doors for critics on the value and ethics of external consultancy.

4. *The pervasive persuasive model*

Over the past 25 years a body of articles [116–119] and books [120–124] –and even television shows [125]– started to question the value, legitimacy and ethics of the management consulting industry. A particular attention has been put on large consulting organizations [121–124]. In this view, consulting is a profit-driven symbolic interaction where either the client or the consultant is a victim. Supporters who put the client as the victim [119–123] report the creation of management fads and fashions that, together with excellent acting skills, persuade corporate executives of the added value. The literature also abounds with examples where it is the consultant (or rather its reputation) which is victim of the client's hidden political agenda, whereby the consultant is brought-in with the sole goal of supporting a predetermined outcome [124, 126]. In this case the credibility of the consulting organization—its cachet—is leveraged to impress shareholders or circumvent resistance from stakeholders. The consultant becomes an extension of top management authorities and so relies on impressions and rhetorics to deliver value.

Other names found in the literature for this model are the *critical model* [48, 103], the *rhetorician* [116], the *impression manager* [120], the *dramaturge* [127], the *creator of management fads* [119]. Of course this model shall not be used as a guide for aspiring consultants: it is an anti-thesis to good management and hence entails none of the successful management theories. I believe however that it is useful for the consultant to be aware of this view. This awareness helps one understand the extreme skepticism that he or she might sometimes face upfront, and what might go wrong in the client-consultant relationship.

The management consulting industry is a relatively new industry: the popular concepts and methods used by consulting organizations were almost all developed within the last few decades. If consultants are to their clients what doctors are to their patients, then consulting will never be accessory nor become obsolete. Recall that in the field of medicine, even the most prestigious organizations had to recognize and eliminate widespread ambiguous practices. Procedures such as invasive cerebral surgeries for example, became famously unethical and absurd only 50 years ago[1] –nowadays it may only be found in Hollywood's productions. And yet no one today would question the legitimacy of neurosurgeons. We might be at an historical juncture when the practice of consulting has to reflect on what is based on science and what is pure impression. But one thing is sure: when supporters of the pervasive model get fame and personal fortune for their achievement, they are neither facilitating this reflection nor helping corporate organizations.

4.1.2 The Models in Practice

The above models are useful theoretical frameworks, but of course different types of relationships will be appropriate depending on the problem circumstances, the project phase and the personalities and incentives of different partners. The client-consultant interaction may also dynamically change during the assignment [48], with mutual dependencies and shifting power relations developing out of simple conversations over problems and solutions [126].

1. *Role of the consultant*

   Arthur Turner identified eight categories of client-consultant interactions [128] that fairly synthesize the categories of models presented above (keeping the *pervasive persuasive* model out…):
   1. Provide information to client
   2. Solve client's problem

[1] The machine learning approach used by Appelbaum et al. [129] is a simple ML regression algorithm introduced in Chap. 7: it consists in assigning a specific weight to each feature and at every step testing the predictive performance, to optimize the set of weights at the next step, effectively learning the relative importance of these features in predicting the overall satisfaction of the client, step by step until the algorithm converges.

3. Make a diagnosis, which may necessitate redefinition of the problem
4. Make recommendations based on the diagnosis
5. Assist with implementation of recommended actions
6. Build consensus and commitment around corrective actions
7. Facilitate client learning
8. Permanently improve organizational effectiveness

Every assignment is unique and in practice the importance of the above categories relative to each other will vary. One could propose that the nature of the client-consultant relationship corresponds to a specific *set of ranking weights* for the set of categories above.

2. *Recent trends*

Recent trends documented in the literature [129] indicate that the relationship is migrating away from the expert model and closer to a partnership agreement characterized by a high degree of mutual trust (toward the *social helper* model, items 6, 7, 8 in Turner's list). Consultancy also migrated further away from a purely strategic exercise to initiate, implement and/or test the recommended course of action (item 5 in Turner's list).

3. *Role of the client*

To first approximation, the client's intervention follows a simple sequence [130]:

1. Select the consulting service provider
2. Explain problem and objectives
3. Provide information to the consultant
4. Check up regularly to follow up on status
5. Evaluate quality of the consultant's intervention

Some consideration should be given to the dual nature of the relationship, as it always involves two types of relations [115]: the cultural dimension (transfer and creation of knowledge) and the political dimension (interest and influence of people). In both dimensions the role played by the client is as critical to the success of an assignment as the role played by the consultant.

4. *Who is the client*

The initial client may be an individual or management team, but as trust and confidence develop between the client and the consultant, consultants begin to see the total organization as "the client" [129]. Edgar Schein [131] recognizes that the issue of who is the client is often ambiguous and problematic. He proposes distinct types of clients, and that the consultant should work on understanding their distinct need, expectation, influence, and degree of participation in the consultancy. There are at least three categories of clients affected by the consultant's intervention (adapted from Ref. [129]):

1. **Primary client**: Individual(s)/unit(s) who contract and manage the intervention
2. **Intermediate client**: All other individual(s)/unit(s) who get involved in the intervention
3. **Unwitting client**: All other individual(s)/unit(s) who may potentially be affected by, but are not involved nor necessarily aware of, the intervention

Some authors [7, 132] note that in some cases it is effective to see the client as the set of interactions and interfaces between individuals/units rather than any specific individual/unit.

5. *The trust issue*

In any case, the role of the client evolves during the relationship and goes beyond the *primary client*. Trust needs be established with everyone who gets involved in the intervention. A common mistake reported in the literature [7] is the "good guy-bad guy" syndrome, where the consultant, by bringing enthusiasm and carrying messiah-like messages, might implicitly signal that one perceives oneself as the good guy, which implies that others are bad guys …or are at least backward.

Both executives and subordinates may experience various concerns and skepticism about any aspect of the consultant's intervention. Without carefully established mutual trust, an individual or unit might feel manipulated and for example keep information for oneself, collude, use information to one's own advantage, criticize colleagues or superiors and give bad reviews during performance evaluation phases [7].

As a general rule of thumb, it was suggested that consultants see themselves as models, willing to practice and develop the effective behaviors that they wish to instill in the client system [7]. As such, consultant's virtues are to give out clear messages, suggest optional methods, encourage and support, give and accept feedbacks in a constructive way, help formulate issues, provide a spirit of inquiry.

6. *Fitness of intervention*

In addition to trust, the consultant's diagnosis and depth of intervention may lead to many adverse effects on the different types of client (*primary*, *intermediate* and *unwitting*). Common mistakes include the temptation to apply a technique that has produced good result in the past but may not be tailored to the client's situation, to intervene at a level deeper than that required for the client's situation, and to recommend action plans incompatible with the client's resources.

In a selfless attempt to avoid the common "good guy-bad guy" syndrome mentioned above, it is tempting to let oneself get seduced into joining the culture of the client organization. Yet, participating in the organization's pathology is likely to neutralize the consultant's effectiveness, not help solve it [7].

7. *The consultant's dilemma*

The current Sect. 4.1 overviewed the different models of client-consultant relationship and how they may apply in practice. The consultant is continuously confronted to the dilemma of whether to "lead and push" or "collaborate and follow". Too close to the expert model and the client tends to be seen as a puppet; too close to the social helper model and it is the consultant who tends to be seen as the puppet. Concerning the persuasive model, it does not provide any insight on how to improve the collaboration between clients and consultants as it views consultants in a pessimistic and critical way. All things considered, the doctor-patient model might be the next best alternative for a simple model to a complex relationship.

The client-consultant relationship extends much beyond the limit of a single assignment. The client's satisfaction determines the likelihood for new assignments and referral to other clients [130]; this long-term relationship is the foundation upon which the entire consulting industry business model is built.

4.2 On the Client's Expectations: Why Hire a Consultant?

The purpose for which a consultant is hired varies depending on the type of consultancy. For example, a case in innovation strategy or IT might require *technical* skills that the client might not possess. But more ubiquitous purposes have also been documented [48]. These include the daily struggle (managerial, political) that managers must cope with and the lack of time for managers to tackle new complex problems when they are already busy with core responsibilities.

Here is a more exhaustive list of "high level" reasons why to hire external consultants, which was provided by McKinsey & Co (adapted from Ref. [129]):

1. Consultants are impartial because they are independent (i.e. Drucker's first value)
2. Consultants have diverse experience outside the client
3. Consultants provide competence not available elsewhere
4. Consultants have time to study the problem
5. Consultants are professionals
6. Consultants have the ability to create action based on their recommendations

The above is a list from the *consultant*'s perspective. Let us now list reasons why to hire external consultants from the *client*'s perspective. The data comes from different surveys of executive managers working in different industries found in the literature (as referenced below):

1. Consultants transfer new competencies to the client company's personal [129]
2. Consultants allow better control on problem solving projects (Sales Representative in Ref. [48])
3. Consultants provide an outsider perspective (CEO in Ref. [48])

4. Consultants can coach and supervise tasks with expert social skills (Sales Director in Ref. [48])
5. Consultants can be used as political weapon [115, 133]
6. Consultants provide additional assistance at no cost –*go the extra mile* [115, 134]
7. Consultants provide customized solutions with concrete implementation planning [129]
8. Consultants provide expert advice [129, 135]

The consultant and client's perspectives overlap well. These insights were derived from diverse surveys including hundreds of professionals, most of whom also expressed drawbacks and challenges not listed here (e.g. in Refs. [48, 129]). Now in order to build both on positive *and* negative feedbacks gathered from large scale clients' surveys, Steven Appelbaum [129] developed an elegant mathematical model of overall client satisfaction from a *client*'s perspective. His team gathered the ratings of 102 officers on more than 50 different aspects of the consultant's interventions. Using the mathematical framework of machine learning described in Chap. 7, where a data set is used to "learn" how well a set of variables (referred to as *independent* variables) may be used to predict the value of another variable (referred to as the *dependent* variable), they created a model that relates overall satisfaction with different aspects of the consultant's intervention[2].

The final result in Appelbaum's study is that a client will give a high rate for the overall project success if, in order of importance:

1. Solutions took into account the client's internal state of readiness
2. Project included prototyping new solutions
3. Project deliverables were clear
4. Consultant partnered with the project team throughout
5. Consultant was professional
6. Consultant understood the client's sense of urgency

This list represents an elegant *data-driven* recommendation toward a successful consulting intervention, as prescribed from the *client*'s perspective.

Earlier in this chapter we noted the importance of both *cultural* and *political* dimensions in the client-consultant relationship. In terms of "expectation", these dimensions underscore the importance of unwritten *psychological* expectations [115] in addition to more concrete *technical* expectations. The technical expectation denotes an in-depth expertise that the client does not a priori possess. The psychological expectation denotes the consultant's overall communication skills and receptiveness to political needs, his/her pragmatism with what can and what cannot be

[2]The strange and curious history of lobotomy did reach a scientific consensus: as put by an anonymous online blogger (below BBC article from H Levinson accessible at www.bbc.co.uk/news/mobile/magazine-15629160): "The science behind it is actually quite solid. The problem comes from trying to cut specific connections, which in the early twentieth century was like trying to destroy an invisible needle in a haystack with a bazooka".

achieved, his/her aptitude to read the environment and fit in the client's team, listen, empathize and provide counsels without charging additional fees [136, 137].

No two consulting projects might ever be the same. Thus, the client cannot know in advance what he/she is exactly buying before he/she gets it [48, 120]. The quality of the interaction, before and during the assignment, will inform the client and calibrate how he/she evaluates the quality of the final deliverable. The take away is that the client's expectation is never completely set in advance. The readiness and quality of the evolving relationship determines the success of a consulting project. It drives the nature of the client's contribution to the solution, and by adjusting his/her expectations governs his/her level of satisfaction.

4.3 Ethical Standards

Management consulting codes of ethics are available through different sources. Most consulting organizations have developed their own codes, but these share similar guidelines. A concise summary is reported below, based on the Association of Management Consulting Firms (AMCF [138]), the Institute of Management Consultants (IMC [139]) and Refs. [7, 130, 140]:

1. *Use of data*
   One shall not distort data to one's advantage nor use it to deceit, punish, or harm.
2. *Professional & technical readiness*
   One shall not distort or misrepresent one's background, capabilities, or expertise.
3. *Confidentiality & conflict of interest*
   One shall not use confidential information from a client to provide competitive advantage to another client, nor disclose information to any group or individual, internal or external, when this information is likely to be used in contradiction with the first rule of consulting.
4. *Coercion & collusion*
   One shall not coerce any individual into disclosing information that they prefer to keep private, nor intentionally collude with some group or individual against any other group or individual.
5. *Openness & promise of realistic outcomes*
   One shall openly communicate the anticipated implications of the proposed course of action.
6. *Disclosure of fees*
   One shall disclose in advance all fee incurred by the recommended intervention, and set forth fees commensurate with the service delivered.

4.4 The First Interview: Defining the Case and Objectives

> *The telephone rings. An executive has some concerns about her organization and the consultant has been recommended as someone who could help. After a brief description of some of the problems and a discussion of the extent to which the consultant's expertise is a reasonable fit for the situation, an agreement is made to pursue the matter over a meal or through an appointment at the executive's office* [7]

The initial phase that leads to a consulting project is notoriously similar across different generations, organizations, problems and industries. It emphasizes the importance of networking and referral systems in the consulting business model [141].

4.4.1 Goals of First Meetings

The first face-to-face meetings aim at exploring the rational and objectives of the client's request. They typically involve a client top executive and a consulting partner, and in subsequent meetings they start to gather *teams* on each respective side.

Some key topics that preface the start of an assignment are *who*, *how*, *when*, and *where* (and for *how much*…).

Who: What group on the client's side will be the starting point of the intervention?

What group on the consultant's side will be a reasonable fit?

How: In what similar circumstances did the client or consultant meet before, how did he/she proceed?

When: What would be an appropriate time frame?

Where: Where would the consultant carry on the different phases of the project?

As pointed out by French and Bell [7], an overriding dimension in these preliminary discussions is the extent of mutual trust that begins to develop between consultant and client. Already at this stage, the aspect of "compensation" entails both *financial* and *psychological* contracts.

4.4.2 Sample of Questions *Consultant-to-Client*

Here is a sample of questions that the consultant might ask to better understand the organization's circumstances (adapted from Ref. [142]):

1. What are your top priorities, what would success look like to you?
2. Where did the problem originate?
3. How ready is the organization for change?
4. What are the current measures in place to solve the problem?
5. What do you think are the root causes of the problem?
6. What are the top challenges that are acting as barriers to success?
7. What actions do you think we should take?

8. Who else should we be talking to?
9. Who will be making the decisions?
10. Is there anything else we should be asking you?

4.4.3 Sample of Questions *Client-to-Consultant* And How to Respond

The client will look for the consultant's ability to adapt to the client's organization landscape, get the ball rolling, and whether he/she has experience in a similar business scenario. Here is a sample of questions that the client might ask (adapted from Ref. [143]):

1. What were your previous client landscapes?
2. What was your role? What was your team's working model?
3. How did you implement a similar business scenario in the past?
4. What was your support structure at your previous client projects?
5. What value did you add at your previous client projects?
6. What best practices did you use before? What were the challenges?
7. How do you keep up with innovation as a consultant?

How to Respond?
The presence of a consultant in an organization automatically instills some psychological expectation from the organization, and the impression that something is needed from its members [115]. The client thus inspects the consultant's *soft* communication skills very closely. Sounding polite and professional, being confident and pragmatic, talking slow, are pre-requisites for the meeting to be successful. If the project involves a specific technology or industry expertise, then the consultant should also convince the organization's members of his/her technical acumen.

Great communication skills and actual content may lead to an excellent relationship, efficient trust building, and collaborative planning. Yet the hallmark of professional consultancy would not be fully introduced if we forgot to talk about the terminology (a.k.a. *buzzwords*) that consultants employ. The words used by consultants are often an efficient marketing tool for brand identification. McKinsey-ites for example refer to their mastery of hypothesis-driven thinking. BCG-ers refer to their proprietary concepts of portfolio matrix and experience curve.

The "creation of consulting fads" is one of the most cited critics against the management consulting industry in the pervasive model[3]. But again, a line should be drawn between what is pure wandering between business concepts that randomly connect with the situation at hand, and what is in contrast a prerequisite for efficient

[3] For a vivid illustration, see Don Cheadle's outstanding acting performance in the first client meeting that takes place in the first episode/first season of the comedy television series House of Lies [125].

communication. Critics should not forget the fundamental reason why experts in any discipline use technical terms at the first place, which is to ease a dialogue by pining down into simple/single words what are otherwise very complex and sometimes not well characterized—yet very real and relevant—concepts.

Management consulting does not enjoy the universal support and centuries-long foundation that most scientific topics enjoy, and discussions are often challenged by the absence of formal unifying support structures, i.e. "scientific" norms and communities. Yet for all its similarities with disciplines such as statistics, psychology and biology, it does engineer discussions and debates as complex as in any of these sciences. In this circumstance, using technical terms *is* a prerequisite. Just as an expert will benefit everyone in a room full of biologists by using technical words such as *fork of replication* instead of digging into an explanation of what it is, a consultant will differentiate from an amateur by using words such as: *landscape analysis, upgrade strategy, disaster recovery, discounted synergies, due diligence, differentiation, diversification, segmentation, disruption, risk mitigation, packaging, re-engineering, scenario planning, blueprinting, breaking even, SWOT, ROI, NPV;* which is just a small (random) sample from a vast colorful ocean of ideas.

Let us close this section with a perspective that is as simple as it is useful to keep in mind: *the first interview with the client is not a job interview*. The client expects the consultant to already possess all relevant technical and functional knowledge to initiate the assignment and perform from day 1.

4.5 Working with the Client *During* the Engagement

The essence of a consulting assignment such as key success factors and rooky mistakes has been discussed in the previous sections. How the overall working style turns out to be depends on the nature of the client-consultant relationship and many other factors. But distinct phases in an assignment can generally be identified independently of the relationship entertained between the two partners. These phases are described below.

4.5.1 Clarifying Objectives

The client and the consultant must come to some understanding on what the desired procedure and outcomes will be. Outcomes are framed into some kind of *measurable* difference for the added value [130]. For example, it could be a predetermined target increase in the bottom line. If everyone is aligned on a measurable target, consultants can more easily get to work.

This stage culminates with a formal proposal from the consultant that delineates some idea of time, scope and fees. In particular, the consultant must demonstrate that he/she understands the cultural and political environment of the client as they relate to the projected outcome of the engagement [115].

4.5.2 Executing Consulting Activities

An engagement implies constant, often day-to-day, interaction with the *primary* and *intermediate* clients [129]. Even for projects that focus on the client external environment, for example understanding the competitive landscape or evaluating the impact of new governmental regulations, it remains essential to understand the culture, politics and needs in the client organization [115]. Indeed the impact of the external environment on the client organization depends on these internal attributes. Hence, in practice, an engagement *always* implies to closely interact with the client. These interactions take the form of group meetings and brainstorming sessions, individual interviews, in person or by phone, and frequent basic exchange of information by emails [144]. The *unwitting* clients come into play indirectly, though these different interactions with *primary* and *intermediate* clients.

A key aspect of the client-consultant interaction involves monitoring and control [145]. This generally takes the form of review meetings with key stakeholders, i.e. *primary* client. At these meetings, consultants present the latest insights and overall progress, their thinking on current issues that relate to the initial objective, and together with the client update on new roadblocks and next course of action [130].

4.5.3 Implementing the Recommendations

Increasingly, clients require that consultants initiate the implementation of their recommendations (see Chaps. 1 and 2). When this is the case, the consultant might replace some members of the team by other professionals who are better equipped to manage specific outcomes. By then, some clients who had remained *unwitting* up-to-now might for the first time interact with, or come under the impact of, the consultant [129].

4.6 Stand and Deliver: Terminating the Assignment

Closing a project versus closing a phase of the project have in common the necessity to obtain formal acceptance by the client: at review meetings, it may be required to obtain a formal authorization to proceed with the next phase of the project lifecycle; at the end of the assignment, it may be required to obtain a formal confirmation that the agreed-upon service has been delivered [130].

At review meetings, the sense of closure is also an opportunity for the consultant to measure client satisfaction (so far), review scope, budget, plans, and ensure access to appropriate resources and infrastructure for the subsequent phases.

The end of the engagement is similar to a review meeting because the end of the project does not imply the end of the relationship [130]. Repeat assignments may benefit both the client and the consultant, financially and management wise. It is indeed often easier to maintain an on-going relationship than to seek and develop a new one. Hence again, the sense of closure is an opportunity for the consultant to ensure client satisfaction, discuss future avenues for collaboration, and document lessons learned from the engagement.

A formal presentation, the *deck*, is often what represents the concretization of the agreed-upon consulting deliverables. The deck should possess certain attributes such as relating back to the objectives, recommending realistic alternative solutions, enabling the client to make informed decisions, and ensuring that the client will be able to proceed with the recommended course of action without the consultant [144].

The two sections below synthetize best practices when preparing and delivering a deck as it relates to *style* and *format*.

4.6.1 Preparing the Slides

1. *Conclusion first – inductive reasoning*

The first slide of the deck is an executive summary. It contains the key insights and conclusions from the consultant's work. It may even already include the recommended course of action.

On several occasions, I asked some management consultants what they believed was the biggest challenge that PhD scientists faced when starting as junior consultant. The answer was always the same! They pointed to *inductive* versus *deductive* reasoning. While deductive reasoning starts with the data and comes to a conclusion, inductive reasoning starts with the conclusion and then rationalizes it based on the data. Scientific communication generally involves deductive storylines where experts interact with other experts on a peer-to-peer basis. Consultants in contrast deliver research insights and recommendations to executive officers who do not possess domain expertise and pay large fees for the delivery.

Inductive reasoning aims to get everybody on the same page upfront, get to the point quickly and avoid losing non-experts in data intensive presentations [144]. Members of the audience are not listening to the consultant's presentation because they connect to the science, but because the consultant will recommend a course of action that is likely to affect their professional and even personal lives. By stating the conclusion upfront the consultant does not instill the suspense that a passionate scientist might strive on, but let members of the audience appreciate what is at stake from the beginning, what is in it for them. It helps them connect to the supporting data whatever their expertise.

An additional –popular– benefit of inductive reasoning is to make your case early (*the elevator ride* [145]) and control how far to go into the details. The deck is the first concretization of a consulting value proposition (i.e. chronologically it always precedes the implementation effort) and thus requires the audience's buy-in [144]. By starting with what is most important and impactful – the conclusions – the consultant can measure buy-in and adapt the depth of discussion to the audience's reaction, whether the audience is receptive to the recommended course of action, indifferent, or actively resistant.

2. *Structure, explain the structure, follow the structure*

The executive summary should be preceded or followed by a table of contents, a.k.a. the *structure* of the presentation. Whether the structure is hypothesis-driven or not, the problem addressed and agreed upon in advance with the client should be the starting point. The *issue tree* that ensues from the original problem statement should be a clear and convincing structure that the audience may easily grasp and follow.

The precise form taken by the issue tree may vary a lot, but it should always build on the original problem statement and follow a MECE (*Mutually Exclusive* and *Collectively Exhaustive*) framework of issues and sub-issues. A method to develop issue trees is introduced in Chap. 5. The rest of the deck flows out of this MECE structure, with different slides addressing different issues and sub-issues. The deck structure enables the consultant to control how far to go into the details depending on the audience's reaction; members of the audience follow this roadmap while the consultant guides them along the way.

Some consultants whom we invited to present a seminar at MIT[4] on the topic of "structuring a deck" proposed the interesting concept of *horizontal and vertical storylines* in the deck. The sequence of slide-titles defines the horizontal dimension. When taken together, these titles should backtrack the main issues of the pre-announced structure. In contrast, the slide-contents are vertical storylines that dig into *sub*-issues and supporting evidences. This concept of horizontal-vertical story-lines provides a simple strategy for how-to adapt to the audience.

3. *Keep it simple*

Conciseness is the hallmark of well-trained consultants and illustrious leaders [146]. The deck should convey ideas to the audience in the clearest, most convincing way possible. The objective of a deck is to communicate a set of recommendations, it is neither a rhetorical nor an acting exercise [144]. Each chart for example will benefit from bringing only one take away.

4. *Use (simple) visuals*

Using an exhibit is an efficient way to argue for a message. For example, the results of a survey could easily be summarized with a simple graphic, while a slide of bullet points would be excessively loaded. A picture is worth a thousand words. If a message may best be constructed with a series of bullet points and nothing to show graphically, it is a sign that it might not show enough of supporting data and give the impression that the message could have been put in a prior memo to save everyone's time [144]. The reason why consultants present the deck orally is because they gather an extraordinary amount of data and boil it down to some key insights. Graphics can be used as a bridge linking the data to the message. Both the message and the link to the data should be presented [144].

[4] Booz Allen Hamilton, MIT 1-day Consulting Workshop on 01/09/2015 (public event).

Each visual will benefit from containing only one message. It is sometimes better to present the same graphic several times and highlight a different piece of information with a distinct message than requesting from your audience to absorb multiple points out of one graphic. As for visual aids (3d plots, animations): the simpler the better. The visual should not come in the way of the message.

Let us close this section with generic best practices: place important information at the top of the slides, use large text font, large images, one graphic and one idea per slide, less than 20 words per slide [144], different types of visuals, high-contrast colors and simple transitions.

5. *Document sources*

Data sources in a deck may take the form of simple pointers (e.g. name of organization that published the data + date of publication) or even small-font footnotes that redirects to an annexed list of sources, but their presence is essential. It is what gives credibility to the consultant and what defines the popular *data-driven* approach. Yet it is surprising how frequently consultants forget to indicate their sources[5]. This is one skill where PhD-trained consultants differentiate, due to the nature of peer-reviewed scientific communication. While doing research for this book, I realized an omnipresence of professionals who entered the consulting industry before graduation and now adopt an all-too narrative communication style: some entire books, including popular non-fiction essays full of recommendations on the data-driven approach (e.g. Victor Cheng's) have been written by consultants without citing any source! As suggested by French and Bell [7], it is advisable that consultants be willing to practice what they preach.

4.6.2 Delivering the Presentation

1. *Follow the structure*

Having prepared a well-structured deck is half of the battle. It demonstrates the consultant's professionalism and empowers his/her confidence. As mentioned earlier, inductive reasoning enables the consultant to measure the audience's buy-in and adapt the level of granularity during the presentation.

For additional generic best practices: write a memory script that features two or three points per slide [144], spend no more than 1 or 2 min per slide, and prepare a few personal or comical stories that will break the ice alongside the deck.

[5]This is not to say that sources should be referenced in any kind of formal (academic) format. But a recurrent theme with prospective consultants is that their presentations completely lack indication of where their data come from. Sources are essential. In term of *format*, using simple pointers (e.g. name of the organization that published the data and date of publication) moved to annexes is a fine practice in consulting presentations.

2. *Avoid surprises*

The client's buy-in is a prerequisite. Building consensus before the presentation will increase the chance that the audience accepts the consultant's recommendations. Indeed, by discussing key insights and conclusions with members of the audience in the intimacy of one-on-one meetings or phone calls, the consultant may receive feedbacks and address potential concerns more easily than in formal group meeting. All these efforts build up support, one at a time, for the actual presentation.

If major decisions have to be made during the meeting, the consultant should even more so seek as much support as possible from the key decision makers before the presentation takes place [144]. Disclosing in advance some information that forces decision makers to change their plans does not secure buy-in but at least it increases the chance that the audience will engage in constructive discussions –and eventually buy in the consultant recommendations—when he/she delivers the presentation.

3. *Adapt to the audience*

Consultant	Let me tell you what I think the problem is
Client	Thanks but I think I understand the problem
Consultant	All right then I don't need to waste your time telling you what your problem is. Let's just turn the first pages over, and we'll go right to the solution

Adapted from Ref. [144]

Being flexible and respectful of the audience is a priority in the client-consultant interaction. As explained earlier, structuring the deck with inductive reasoning is an effective approach to adapt the depth of discussion to the audience's reaction, whether this reaction is supportive, indifferent, or actively resistant. Adapting to the audience also requires understanding expectations, goals, backgrounds, preferred styles and languages. As he/she delivers the presentation, the consultant may highlight different aspects of the deck's structure and adapt style and languages. Below is a list of audience attributes that might be useful for the consultant to consider prior to a client meeting:

1. **Expectation**: primary client, intermediate client, financial manager, operation manager
2. **Background**: manager, executive, scientist, specialist, salesman, IT personnel
3. **Style**: formal vs. informal, technical vs. simplified, intimate vs. detached
4. **Language**: blue collars vs. white collars, state of mind, preferred terminologies.

The Structure of Consulting Cases

5

5.1 How to Develop a Tailored Structure?

A dilemma

A classic dilemma for consulting practitioners is how much to rely on pre-defined frameworks. When does a structure stop qualifying as *one size fits all* and start *reinventing the wheel*? How to strike the right balance?[1]

Developing a tailored case structure is a frequent source of failure from candidates at case interviews because there is no clear consensus on what represents a good *MECE* (mutually exclusive, collectively exhaustive) framework [88]. Some consultants openly recommend *never* using preexisting frameworks, yet developing multi-prong, structured approaches. I, for one, received such advice from a consultant at a McKinsey social gathering [147]. Others in contrast, consider reference structures and experience to be necessary building blocks when developing a tailored case structure. Who is to say who is right?

At least for the purpose of this chapter, let us assume that reference frameworks such as the ones presented in Sects. 5.2, 5.3, 5.4, 5.5, 5.6, 5.7 and 5.8 *should be* leveraged when developing a tailored case structure. This is because many researchers, including yours truly, believe that other types of discourse suffer from pedagogic inconsistencies and learning problems in practice: it may be effective to follow your gut and be creative when you have been in the industry for a few years, but for a beginner to adopt best practices, leveraging the widely popular frameworks certainly is useful. The current section provides recommendations for tailoring, and Sects. 5.2, 5.3, 5.4, 5.5, 5.6, 5.7 and 5.8 a set of high-level thinking roadmaps that, to the best of my knowledge, have met success in the past indeed.

---

[1] A related discussion on the *wave of commoditization* brought about by the emergence of big data in management consulting can be found in Chap. 1 (Sect. 1.3.2).

J. D. Curuksu, *Data Driven*, Management for Professionals,
https://doi.org/10.1007/978-3-319-70229-2_5

Tailoring

The concept of inductive reasoning introduced in Chap. 4 (Sect. 4.6.1) is key to develop a tailored case structure. A tailored case structure corresponds to an issue-tree, and may be constructed by following five steps:

Step 1: *Define one key focal issue*

This initial issue should be carefully designed with the client.

Step 2: *Brainstorm key sub-issues*

The second step is where tailoring already starts. A classic mistake is to skip steps 2–4 and directly use a popular framework along the lines of those suggested in Sects. 5.2, 5.3, 5.4, 5.5, 5.6, 5.7 and 5.8. Indeed, the framework should be so developed that all sub-issues considered relate to the key focal issue in a way that the consultant can easily "articulate", i.e. how it relates to the key focal issue should become perfectly clear when presented to the client. But again, the consultant needs to strike the right balance: to efficiently brainstorm and navigate through relevant ideas, the popular frameworks and his/her personal experiences will both be of assistance.

Step 3: *Build an issue tree by clustering sub-issues into MECE categories*

Step 3 consists in cleaning the set of issues gathered in step 2. Goal is to avoid redundancies and clarify meanings. Steps 2 and 3 are recursive since step 3 might help detect missing issues and send us back to step 2.

Step 4: *Prioritize issues and sub-issues*

Step 4 consists in rearranging the set of issues selected in step 3, by prioritizing the most important issues for which data can be gathered and analyzed within a reasonable time frame.

Step 5: *Ask questions to gather facts and (in-) validate hypotheses*

Step 5, gathering facts and figures, is where the research part of the assignment really starts. The structure developed in steps 2 to 4 provides pointers to delegate tasks in the team and assess progress in the project.

Hypothesis generation

The key focal issue as well as the sub-issues are often called hypotheses, which is a more direct form of inductive reasoning whereby the researcher starts with an intuitive answer based on its experience or expertise, and gather facts to (in-) validate the hypothesis. Hypothesis-generation is often used because it leads to more focused and thus potentially more efficient data gathering.

Client bias

When developing a structure, it is also important to resist the temptation of following the client's suggestions too closely [144]. A key feature explaining why management consulting met so much success in the past is that diagnosing a problem and identifying root causes both require unbiased and analytical inquiries. Even if business experience and acumen may help prioritize some lines of inquiries, a consulting diagnostic and prescription ultimately must rely on facts and figures.

In the next Sects. (5.2, 5.3, 5.4, 5.5, 5.6, 5.7 and 5.8), a set of frameworks is provided to help consultants navigate the range of high level issues that may alight upon their cases. The proposition for a "*one-size-fits-all*" was designed based on Refs. [18, 19, 46, 100, 148–150] and personal experiences, the *profit* framework, *pricing* framework and *mergers and acquisitions* framework were adapted from Refs. [100, 148–150], the *operation* framework and *growth and innovation* framework were designed based on Refs. [18, 19, 46, 47, 100] and the *new ventures and startups* framework was designed based on Refs. [18, 47, 100, 151, 152], plus personal experiences.

5.2 Proposition for a "*One-Size-Fits-All*"[2]

Table 5.1 Proposition for a "one-size-fits-all"

Business Case		
Industry	Competition	Client
Growth	**Shares**	**Capabilities**
Customers:	**Basis of competition**	**Finances**
- Segments	**Prices**	**Products:**
- Characteristics	**Competitive response**	- Segments
- Jobs to be done	**New entrants**	- Differentiation
- Profitability	**Trends**	- Profitability
- Trends		- Trends
Barriers to entry		**P&L**
Risks		**Best practices**
		Value chain

1. *Industry*
   - How is the industry doing? How is the client growing compared to industry average?
   - Define customer segments (e.g. by demographics, psychographics or *jobs*[3]). What are each segment's characteristic needs/problems to solve?
   - How well is the client helping with customer jobs to be done in each segment?
   - Which segments are most profitable? How did they evolve recently?
   - What are the key barriers to entry? (capital requirement, regulations, access to distribution, IP)
2. *Competition*
   - What are the market shares? How did they evolve recently?
   - On what is competition mainly based?
   - Are the client's prices in line with competitors?
   - What will be the competitive response?
   - Have there been new entrants lately?

[2] This is also referred to as "*one-size-fits-none*", so beware!

[3] The notion of jobs-to-be-done is introduced in Refs. [18, 19, 47] and described in Sect. 3.2.

- Have some competitors changed recently? (e.g. price, product, marketing, strategy)

3. *Client*
   - What are the client's capabilities and disabilities? (e.g. resources, processes, values)
   - How is the client doing financially? Are there cash reserves?
   - What are the different products? How do the products work?
   - How does the client's products differentiate from competition?
   - Which products are the most profitable?
   - How did the different business units evolve recently? Which products have the most potential?
   - What are costs and revenues? How did they evolve compared to competitors?
   - How does the value chain work? Have there been issues with operation, supply, distribution?

5.3 The Profit Framework

Table 5.2 The profit framework

Profit issue			
Scope	Revenues	Costs	Custom tree
Industry	**Revenue streams:**	**Cost breakdown:**	**Growth**
Why:	- Price per unit	- Fixed costs	**Organization**
- Goals	- Sales volume	- Cost per unit	**Due diligence**
What, how:	**Percentages**	- Production volume	**Others**
- Products	**Trends Balance**	**Trends**	
- Business model		**Balance**	

1. *Scope*
   - How is the industry doing? How is the client growing compared to industry average?
   - What are the market shares? How did they evolve recently?
   - What are the objectives?
   - What are the products?
   - How does the client's business model work and differentiate in the market place?

2. *Revenues*
   - What are the revenue streams?
   - Define segments (e.g. by product line, customer segment, distribution channel, or geography)
   - For each segment: what are the price per unit and sales volumes?
   - What percentage of total revenue does each stream represent? (look for high revenue products)
   - How have revenue streams and percentages changed lately? Why did these changes take place? Does anything seem unusual?

3. *Costs*
   - How are the costs distributed?
   - What are the fixed costs and investments?
   - What are the variable cost per unit and production volumes?
   - How have costs changed lately? Why did these changes take place? Does anything seem unusual?
4. *Custom tree*

The above analysis delivers a quantitative explanation for the client's profit problem, and spells out potential root causes. Based on this diagnostic, the assignment may then advance to a more qualitative stage of inquiry into the internal and/or external circumstances responsible for the observed symptoms, and in the pursuit of tailored solutions.

5.4 The Pricing Framework

Table 5.3 The pricing framework

Pricing Strategy			
Scope	Costs analysis	Competitive analysis	Economic analysis
Objective: - Short-term - Long-term **Level of control:** - Internal strategy - Reaction to: - Suppliers - Market - Competitors	**Variable costs** **Fixed costs** **Breakeven analysis**	**Differentiation** **Prices** **Costs** **Strategies** **Response of competitors**	**Jobs to be done** **Willingness of market to pay** **Market sizing**

1. *Scope*
   - Is the client motivated by short-term return (increasing profit ASAP) or long-term return (e.g. increasing market shares)?
   - Does the case arise from an internal growth strategy initiative, or is the client reacting to recent external events?
2. *Cost analysis*
   - What are the variable costs, fixed costs and investments?
   - How have costs changed lately? And why?
   - What is the breakeven volume for different price points?
3. *Competitive analysis*
   - How does the product compare with competitive offerings and substitutes?
   - What are the prices of competitive offerings and substitutes?
   - Do we have information on their costs? How about their pricing strategies?
   - Think about how competitors will react to our proposed strategy.

4. *Economic analysis*
    - What customer jobs to be done does the product address? What problems is it trying to solve?
    - Think about how much customers would be willing to pay for this product.
    - What is the total market size? Can we identify potential non-consumption or currently over-served market opportunities? (e.g. disruptive innovations [18] or blue oceans [77], see Chap. 8 for details)
    - Which phase of its growth cycle is the industry in?

5.5 Operations

Table 5.4 Operations

Operational Issue			
Business model	Value chain	Benchmarks	Action plan
Industry: - Trends - Value network - Regulations **Portfolio/SBUs** **Differentiation:** - Competition basis - Market shares **Capabilities**	**Procurement:** - Suppliers transportation - Storage/handling **Processing:** - Procedure Technology - Maintenance **Delivery** (e.g. Packaging) **Marketing:** - Promotions - Sales channels - Customer support	**Competition:** - Costs per unit - Volumes - Cost effectiveness - Trends **Innovations:** - Technologies - Product features - Business models	**Milestones:** - Control systems - Innovation re-engineering - M&A - Budgeting - Cost reduction - Downsizing

1. *Business model*
    - How is the industry doing overall? How is the client doing compared to the industry?
    - What customer jobs to be done do the client's products address? What problems are they trying to solve?
    - Are there important regulations in this industry?
    - Overview strategic business units and product portfolio.
    - How does the client compete in the market place? (differentiation, cost leadership, focus)
    - What are the market shares? How did they evolve recently?
    - What are the client's capabilities and disabilities? (resources, processes, values)

2. *Value chain*
   - How do procurement, processing, delivery and marketing work?
   - What are the costs per unit for each activity of the value chain?
   - How cost-effective are current promotional tools and sales channels?
3. *Benchmarks*
   - How does each activity of the value chain compare with competitors? (costs, effectiveness)
   - Can we find innovative ideas in-/out-side the organization that have potential for its business?
4. *Action plan*

The above analysis delivers a set of opportunities for improving operations. Based on these insights, the assignment can then advance to a more action-oriented stage of inquiry, into the innovations that have potential to maximize value and minimize cost for each operational activity.

5.6 Growth and Innovation

Table 5.5 Growth and innovation

Growth Strategy			
Market opportunities	Client capabilities	Menu of strategies	Action plan
Landscape: - Trends - Success factors - Regulations **Customers:** - Segments - Potentials - Market sizes **Competition:** - Shares - Competition basis - Prices	**Differentiation** **Portfolio/SBUs** **Operations** **Innovations** **Finances**	**Options:** - In-/decrease prices - Price dynamically - Change product line - Change packaging - Change serving sizes - Change distribution - Increase services - Increase marketing - Innovate product - Innovate model - Reengineer model - Create alliances - Acquire competitors - Diversify	**Management:** - Subsidiary acquisition - Alliance **Milestones:** - Budgets - Structures - Policies - IT systems - Training systems - Control systems **Risk analysis**

1. *Market opportunities*
   - How is the industry doing (growth cycle)? How is the client doing compared to the industry?
   - What are the key success factors in this industry?

- Are there important regulations in this industry?
- Define customer segments (demo-/psycho-graphics, jobs); Which segments are the most profitable?
- What do customers look-for in each segment, what problems are they trying to solve?
- Are there potential non-consumption or currently over-served consumption opportunities?
- Can we find innovative ideas in-/out-side the organization that have potential for the business?
- What are the market sizes?
- What are the market shares? How did they evolve recently?
- On what competition is mainly based? Are the client's prices in line with competitors?

2. *Capabilities*
   - What are the client's core competencies and competitive edges? (resources, processes, networks)
   - What is the client's current product portfolio? Which products have the most potential?
   - How does the client operate its value chain? (procurement, processing, delivery, marketing)
   - How advanced is the client's innovation culture? (teams, structures, systems)
   - How is the client doing financially? Are there cash reserves?

3. *Menu of strategies*

What strategic direction(s) best fit the situation? The *Ansoff* matrix [153] in Fig. 8.4 offers a high-level thinking roadmap: penetration, market vs. product development, or diversification.

4. *Action plan*

The above analysis delivers a set of opportunities for growth and innovation. Based on these insights, the assignment can then advance to a more action-oriented stage of inquiry: determine what division or organization should execute the strategy, develop mechanisms to manage the interface with the organization, and analyze risks based on alternative scenarios such as new technology, regulation, market trends and competitive response.

5.7 Mergers and Acquisitions

Table 5.6 Mergers and acquisitions

Mergers and Acquisitions				
	Due diligence			
Scope	Client	Target	Together	Outlook
Motivations:	**Products**	**Products**	**Products**	**Risks:**
- Market share	**Customers**	**Customers**	**Customers**	- Legal
- Synergies	**Value chain:**	**Value chain:**	**Value chain:**	- Regulatory
- Diversify	- Procurement	- Procurement	- Procurement	- Technology
- Shareholders	- Operations	- Operations	- Operations	- Competitive response
- Taxes	- Distribution	- Distribution	- Distribution	**Milestones:**
- Reselling	- Marketing	- Marketing	- Marketing	- Form a special unit
Industry:	**Culture**	**Culture**	**Culture**	- Get support/ commitment to one single culture
- Jobs	**Management**	**Management**	**Management**	
- Players	**P & L**	**P & L**	**P & L**	
- Shares	**Valuation** (e.g. P/E ratio, NPV)			
- Trends				
- Regulations				

1. *Scope*
   - What are the objectives? (market share, synergies, product diversification, competitive response, stock value, tax advantages, break target up and sell off parts)
   - What customer jobs to be done do the client's products address? What problems are they trying to solve?
   - How is the industry doing overall? How is the client doing compared to the industry?
   - What are the market shares? How did they evolve recently?
   - Are there important regulations in this industry?

2. *Due diligence*

   Compare the potential of business models between the client and the potential target(s), and estimate the potential of a combined model:
   - What are the product offers, customer bases, market shares, value chains and cultures?
   - Are there synergy opportunities to add value through better management?
   - Estimate revenues, costs, synergy opportunities and resulting profits.
   - Value the target using a quick Price-Earning ratio estimate or a more rational Net Present Value computation that factors in discounted cash flows, discounted synergies and perpetuities [149]

3. *Risks and implementation*

Are there any legal reasons, governmental regulation, or technology risks why the client should not move forward with integrating the proposed target company? How will competitors react?

Form a separate team that will oversee the integration, get support and contribution from both partners, and develop incentive programs to encourage employees to commit to only one culture.

5.8 New Ventures and Startups

Table 5.7 New ventures and startups

New Ventures			
Business model	Management	Customers	Action plan
Landscape:	**Capabilities**	**Segments:**	**Refine the business model**
- Value network	**Finances**	- Characteristics	**Engage with customers**
- Competition	**Current**	- Potentials	**Milestones:**
- Success factors	**Portfolio:**	- Trends	- Fund gathering
- Barriers to entry	- Potentials	**Innovations:**	- Budgeting
Entry strategy	- Differentiation	- Technologies	- Partnership
Operations	- Value chain	- Product features	- Control systems
Pricing	**Fit**	- Business models	
	P&L forecasts	**Market sizing**	

1. *Business model*
    - What customer jobs to be done, do the client's products address? What problems are they trying to solve?
    - How will the client compete in the market place? (differentiation, cost leadership, focus)
    - How did market shares for key players evolve recently, what are the key success factors?
    - What are the key barriers to entry? (e.g. resources, regulations, access to distribution, IP)
    - What are the advantages/disadvantages of starting from scratch vs. acquisition vs. joint venture?
    - How will the client operate its value chain (procurement, processing, delivery, marketing)
    - How will the client price the product?
2. *Management*
    - What are the accreditations and competencies of the management team?
    - How is the client doing financially? Are there cash reserves?
    - What capital structure (debt vs. equity) and allocation of funds may be considered?

- What is the client's current product portfolio? Which products have the most potential?
- How does the client currently differentiate from competition?
- Will the current operational workflow benefit in/from the new business?
- How will the new business fit with the rest of the product portfolio? (synergy vs. cannibalization)
- Forecast costs and revenues to compute the expected ROI for different time horizons

3. *Customers*
   - Define customer segments (demo-/psycho-graphics, jobs); Which will likely be the most profitable?
   - Can we find innovative ideas in-/out-side the organization that have potential for the business?
   - What is the total market size? Can we identify potential non-consumption or currently over-served market opportunities? (e.g. disruptive innovations [18] or blue oceans [77], see Chap. 8 for more details)

4. *Action plan*

The above analysis delivers a better understanding of customers, competitors, innovators, and the overall potential of the client's new venture. Based on these insights, the assignment can then advance to a refinement and implementation stage: integrate new features in the business model by reviewing in turn the resources, processes and values, engage with customers through surveys and promotional campaigns (e.g. free trials, demonstrations) and articulate milestones that will bolster the integration of the value proposition in the overall industry value *network*.

Principles of Data Science: Primer

6

Let us face it. Statistics and mathematics deter almost everyone except the ones who choose to specialize in it. If you kept reading and reached this far in the book you are probably now considering skipping the chapters on *Data Science* and moving on to the next on *Strategy* because, well, it sounds more exciting. Thus, let us start this chapter on statistics by a simple example that illustrates why it is worth reading and why consultants may increasingly use mathematics.

Suppose you gathered information on the demographics and psychographics of thousands of customers with extensive surveys. Suppose the data includes numerical variables such as age, income and mortgages, plus categorical variables such as education, health problems and travel preferences. Your client provided you with all transaction records made by credit card over the past 5 years. In order to think about your client's growth options, you would like to answer a simple question: who are the best customers? The [big] data is here, you have intelligence on tens of thousands of transactions themselves associated with particular income levels, emotional preferences, etc. You could barely hope for more data. But you start to wonder… what does "best customer" mean? Seemingly the ones who purchased the most in the database are potential candidates. But in your effort to understand these customers, you start to further wonder what are the characteristics of these customers? What *features* correlate or associate with purchase levels? Does education background for example have anything to do with purchase levels? Maybe not, but in the category of high-income customers, now, does it? Can one make better predictions by using only the income levels, or will a combination of income levels *and* education background deliver better results? At the end of the day, what subset of features might best predict purchase levels and represent a good starting point for your team to brainstorm strategic options? How confident may you be in all these predictions?

The above problem is as simple as it is important – it asks: what features have something to do with purchase and by how much. But this problem is impossible to solve mentally. Some data are missing (purchases made in cash), some data cannot be directly compared (continuous variables versus binary yes/no answers), and

J. D. Curuksu, *Data Driven*, Management for Professionals,
https://doi.org/10.1007/978-3-319-70229-2_6

some data will turn irrelevant. And here is why data science is worth learning. With the mathematical framework of *machine learning* [89] introduced in Sect 6.2, if there exists a set of say three features in this database that correlate with purchase levels, with degree of confidence >95% (p-value <0.05), it will be possible to find these three features within just a few minutes on a computer. This is what data science has to offer. Once one understands the basics, what no rational thinking could ever hope to accomplish even in a lifetime may be delivered by a computer within the hour.

This Primer

This primer is not an introduction to statistics, which would require at least an entire book[1]. The goal of the data science chapters is to overview the basic mathematical tools and concepts that may be used in business management, some typical workflows, and some applications. Common sense and focus are certainly required in these chapters at least as much as in any other. But armed with this commitment (of focus—we will assume everyone possesses common sense…), there is indeed nothing in these chapters that shall be out of reach intellectually. It shall, however, increase your confidence when you analyze data in the future.

A typical data analysis project starts with exploring and cleaning the data, then moves on developing theories to interpret the data, and ends with communicating takeaways and/or predictions based on intelligence gathered from the data. What happens next, be it either *discussions* with stakeholders or *applications* of the model to new data, both represent an ultimate *refinement* phase where theory meets practice, where the model interacts further with real-world data. Indeed, acceptance vs. resistance from stakeholders and success vs. failure of an application to new data, may both lead to valuable knowledge and refinement. This real-world feedback is increasingly leveraged in most data analysis projects due to the emergence of *big data* and the recent revolution in the economic of information, as discussed later in this chapter.

A note on equations

In the data science chapters the key equations are provided because, as the popular saying has it for pictures and graphs, an equation may be worth a thousand words. But each equation has been stripped to its minimum formulation, and no technical background is assumed. Thus with a fair amount of focus and commitment, the following material is accessible to all readers irrespective of their left-brain/right-brain inclination. An exception might be the repeated use of the *integral* symbol $\int$ where the reader could expect the *sum* symbol $\sum$, so let us get that out of the way: $\int$ and $\sum$ are completely equivalent for the purpose of this discussion and $\int$ was chosen only because it is visually more elegant than $\sum$ when repeated multiple times. A formal difference exists ($\int$ applies to continuous variables, $\sum$ to discrete variables) but is irrelevant for the purpose of an introduction to data science.

[1] One may recommend "*Naked Statistics*" from Charles Wheelan [89], which introduces the overall field of statistics in a simple and humoristic way …technical expertise not required.

6.1 Basic Mathematic Tools and Concepts

Descriptive Statistics

The process of analyzing data generally starts in the form of *descriptive* statistics where one tries to understand and summarize the information contained in the data-set [91]. Eventually, the analysis steps forward to *inferential* statistics, where probability theory (Sect. 6.2) is used to extrapolate meanings from *sample* to *population* and make effective predictions.

Consultants perform in the field of descriptive statistics all the time. When they use measures of central tendency such as the mean, median and modes, or measures of variability such as the maximum, minimum, variance, standard deviation and quintiles, they are performing in the field of descriptive statistics.

Descriptive statistics also includes the initial step of *data exploration*, which is what consultants do when they use visualization tools and graphical illustrations such as scatter plots (i.e. raw data on a graph), density curves and histograms (used to visualize proportions and percentages), box plots, circular diagrams, etc.

Finally, to summarize and advance one's understanding of the information contained in a dataset, it is necessary to rely on some basic theoretical underpinnings, such as recognizing a particular distribution function (e.g. the Normal "bell curve", a Binomial, an Exponential, a Logarithmic, a Poisson, a Bernoulli), defining outsiders when beyond one or several standard deviations from the mean, looking at cumulative distributions (i.e. probability to be within a *range* of outcomes), etc.

Cross Validation

Convergence of the sampled data needs to be assessed. This is often referred to as *robustness*, i.e. how valid the information that one may extract from the data is when moving from one sample to another. The key challenge to assess convergence is that multiple samples need to be available, and this is generally not the case! A common approach employed to solve this problem is to subdivide the sample into sub-samples, for example using *k-fold* cross-validation [154] where k is the number of sub-samples. An optimal value of k may be chosen such that every sub-sample is randomly and homogeneously extracted from the original sample, and the convergence is assessed by measuring a *standard error*. The standard error is the standard deviation of the means of different sub-samples drawn from the original sample or population.

The method of k-folding has broad and universal appeal. Beyond the simple convergence checks that it enables, it has become key to contemporary statistical learning (a.k.a. *machine learning*, see examples and applications in Sects. 6.2 and 6.3). In statistical learning, some learning sets and some testing sets are drawn from a "working" set (the original sample or population). Most common forms of k-folding are 70% hold-out where 70% of the data is used to develop a model and then 30% is used to test the model, 10-fold cross validation where nine out of ten sub-samples are recursively used to train the model, or n-fold (leave one out) iterations where all but one data point is recursively used to train the model [154].

The above methods enable to seek the most robust model, i.e. the one that provides the highest score across the overall working set by taking advantage of as much information as possible in this set. In so doing, k-folding enables the so-called learning process to take place. It integrates new data in the model definition process, and eventually may do so in real-time, perpetually, by a robot whose software evolves as new data is integrated and gathered by hardware devices[2].

Correlations

The degree to which two variables change together is the *covariance*, which may be obtained by taking the mean product of their deviations from their respective means:

$$\text{cov}(x,y) = \frac{1}{n}\int_{i=1}^{n}(x_i - \bar{x})(y_i - \bar{y}) \tag{6.1}$$

The magnitude of the covariance is difficult to interpret because it is expressed in a unit that is literally the product of the two variables' respective units. In practice thus, the covariance may be normalized by the product of the two variables' respective standard deviations, which is what defines a correlation according to Pearson[3] [155].

$$\rho(x,y) = \frac{\text{cov}(x,y)}{\frac{1}{n}\sqrt{\int_{i=1}^{n}(x_i - \bar{x})^2 \int_{i=1}^{n}(y_i - \bar{y})^2}} \tag{6.2}$$

The magnitude of a correlation coefficient is easy to interpret because −1, 0 and +1 can conveniently be used as references: the mean product of the deviation from the mean of two variables which fluctuate in the exact same way is equal to the product of their standard deviations, in which case $\rho = +1$. If their fluctuations perfectly cancel each other, then $\rho = -1$. Finally if for any given fluctuation of one variable the other variable fluctuates perfectly randomly around its mean, then the mean product of the deviation from the means of these variables equals 0 and thus the ratio in Eq. 6.2 equals 0 too.

The intuitive notion of correlation between two variables is a simple *marginal* correlation [155]. But as noted earlier (Chap. 3, Sect. 3.2.3), the relationship between two variables x and y might be influenced by their mutual association with a third variable z, in which case the correlation of x with y does not necessarily imply causation. The correlation of x with y itself might vary as a function of z. If this is the case, the "actual" correlation between x and y is called a *partial correlation* between x and y given z, and its computation requires to know the correlation between x and z and the correlation between y and z:

[2] The software-hardware interface defines the field of *Robotics* as an application of *Cybernetics*, a field invented by the late Norbert Wiener and from where *Machine Learning* emerged as a subfield.

[3] *Pearson correlation* is the most common in loose usage.

$$\rho(x,y)_z = \frac{\rho(x,y) - \rho(x,z)\rho(z,y)}{\sqrt{1-\rho^2(x,z)}\sqrt{1-\rho^2(z,y)}} \tag{6.3}$$

Of course in practice the correlations with z are generally not known. It can still be informative to compute the marginal correlations to filter out hypotheses, such as the presence or absence of strong correlation, but additional analytics techniques such as the regression and machine learning techniques presented in Sect. 6.2 are required to assess the relative importance of these two variables in a multivariable environment. For now let us just acknowledge the existence of partial correlations and the possible fallacy of jumping to conclusions that a marginal correlation theoretical framework does not actually offer. For a complete story, the consultant needs to be familiar with modeling techniques that go beyond a simple primer so we will defer the discussion to Sect. 6.2 and a concrete application example to Sect. 6.3.

Associations

Other types of correlation and association measures in common use for general purposes[4] are worth mentioning already in this primer because they extend the value of looking at simple correlations to a broad set of contexts, not only to quantitative and ordered variables (which is the case for the correlation coefficient ρ).

The *Mutual Information* [156] measures the degree of association between two variables. It can be applied to quantitative ordered variables as for ρ, but also to any kind of discrete variables, objects or probability distributions:

$$MI(x,y) = \iint_{x,y} p(x,y)\, \log\left(\frac{p(x,y)}{p(x)p(y)}\right) \tag{6.4}$$

The *Kullback-Leibler* divergence [157] measures the association between two sets of variables, where each set is represented by a multivariable (a.k.a. *multivariate*) probability distribution. Given two sets of variables $(x_1, x_2, \ldots, x_n)$ and $(x_{n+1}, x_{n+2}, \ldots, x_{2n})$ with multivariate probability functions p_1 and p_2, the degree of association between the two functions is[5]:

$$D(p_1, p_2) = \frac{1}{2}\left(cov_1 : cov_2 - ln\left(\frac{\det(cov_2^{-1})}{\det(cov_1^{-1})}\right) - I:I\right) + \frac{1}{2}\left(cov_2^{-1}(\bar{\mathbb{x}}_1 - \bar{\mathbb{x}}_2)^2\right) \tag{6.5}$$

where a colon denotes the standard Euclidean inner product for square matrices, $\bar{\mathbb{x}}_1$ and $\bar{\mathbb{x}}_2$ denote the vector of means for each set of variables, and I denotes the identity matrix of same dimension as cov_1 and cov_2, i.e. $n \times n$. The Kullback-Leibler divergence is particularly useful in practice because it enables a clean way to

[4] By general purpose, I mean the assumption of linear relationship between variables, which is often what is meant by a "simple" model in mathematics.

[5] Eq. 6.5 is formally the divergence of p_2 from p_1. An unbiased degree of association according to Kullback and Leibler [157] is obtained by taking the sum of each one-sided divergence: $D(p_1,p_2) + D(p_2,p_1)$.

combine variables with different units and meanings into subsets and look at the association between subsets of variables rather than between individual variables. It may uncover patterns that remain hidden when using more straightforward, one-to-one measures such as the Pearson correlation, due to the potential existence of partial correlation mentioned earlier.

Regressions

The so-called Euclidean geometry encompasses most geometrical concepts useful to the business world. So there is no need to discuss the difference between this familiar geometry and less familiar ones, for example curved-space geometry or flat-space-time geometry, but it is useful to be aware of their existence and appreciate why the familiar *Euclidean distance* is just a concept after all [158], and a truly universal one, when comparing points in space and time. The distance between two points x_1 and x_2 in a n-dimensional Euclidean Cartesian space is defined as follows:

$$d = \sqrt{\int_{i=1}^{n} \left(x_{i1} - x_{i2}\right)^2} \tag{6.6}$$

where $2D \Rightarrow n = 2$, $3D \Rightarrow n = 3$, etc. Note that $n > 3$ is not relevant when comparing two points in a real-world *physical* space, but is frequent when comparing two points in a *multivariable* space, i.e. a dataset where each point (e.g. a person) is represented by more than three features (e.g. age, sex, race, income $\Rightarrow n = 4$). The strength of using algebraic equations is that they apply in the same way in three dimensions as in n dimensions where n is large.

A method known as *Least Square approximation* which dates back from late eighteenth century [159] derives naturally from Eq. 6.6, and is a great and simple starting point to fitting a model to a cloud of data points. Let us start by imagining a line in a 2-dimensional space and some data points around it. Each data point in the cloud is located at a specific distance from every point on the line which, for each pair of points, is given by Eq. 6.6. The shortest path to the line from a point A in the cloud is unique and orthogonal to the line, and of course, corresponds to a unique point on the line. This unique point on the line *minimizes d* since all other points on the line are located at a greater distance from A, and for this reason this point is called the *least square solution* of A on the line. The least square approximation method is thus a minimization problem, the decisive factor of which is the set of square differences of coordinates (Eq. 6.6) between observed value (point in the cloud) and projected value (projection on the line), called *residuals*.

In the example above, the line is a *model* because it projects every data point in the cloud, a complex object that may eventually be defined by a high number of equations (which can be as high as the number of data points itself!), onto an simpler object, the line, which requires only one equation:

$$x_2 = a_1 x_1 + a_0 \tag{6.7}$$

The level of complexity will be *good enough* if the information lost in the process may be considered *noise* around some kind of background state (the *signal*).

The fit of the model to the data (in the least square sense) may be measured by the so-called *coefficient of determination* R^2 [160, 161], a signal-to-noise ratio that relates variation in the model to variation in the data:

$$R^2 = 1 - \frac{\int_{i=1}^{k} \left(x_{2ob}^{(i)} - x_{2pr}^{(i)} \right)}{\int_{i=1}^{k} \left(x_{2ob}^{(i)} - x_{2av} \right)} \tag{6.8}$$

where x_{2ob} is an observed value of x_2, x_{2pr} is its estimated value as given by Eq. 6.7, and x_{2av} is the observed average of x_2. k is the total number of *observations*, not to be confused with the number of dimensions n that appears in Eq. 6.6. R^2 is a sum of differences between observed versus predicted values, all of which are *scalars*[6], *i.e.* values of only one dimension. R^2 is thus also a scalar.

To quickly grasp the idea behind the R^2 ratio, note that the numerator is the sum of residuals between observed and predicted values, and the denominator is the variance of the observed values. Thus, R^2 tells us *what percent of the variation in the data is explained by the regression equation.*

The least-square modeling exercise above is an example of linear regression in two dimensions because two variables are considered. Linear regression naturally extends to any number of variables. This is referred to as *multiple regression* [162]. The linear least-square solution to a cloud of data points in 3 dimensions (3 variables) is a *plane*, and in n dimensions (n variables) a *hyperplane*[7]. These *generalized linear models* for regression with arbitrary value of n take the following form:

$$x_n = a_0 + a_1 x_1 + a_2 x_2 + \ldots + a_{(n-1)} x_{(n-1)} \tag{6.9}$$

The *coefficients* a_k are scalar parameters, the *independent features* x_i are vectors of observations, and the *dependent response* x_n is the predicted vector (also called *label*). Note that the dimension of the model (i.e. the hyperplane) is always $n-1$ in a n-dimension space because it expresses one variable (*dependent label*) as a function of all other variables (*independent features*). In the *2D* example where $n = 2$, Eq. 6.9 becomes identical to Eq. 6.7.

When attempting to predict several variables then the number of dimensions covered by the set of features of the model decreases accordingly. This type of regression where the response is multidimensional is referred to as *multivariate regression* [163].

By definition, hyperplanes may be modeled by a single equation which is a weighted product of first powers of $(n-1)$ variables as in Eq. 6.9, without

[6] All 1-dimentional values in mathematics are referred to as *scalars*; multi-dimensional objects may bear different names, most common of which are *vectors*, *matrices* and *tensors*.

[7] *Hyperspace* is the name given to a space made of more than three dimensions (i.e. three variables). A plane that lies in a hyperspace is defined by more than two vectors, and called a *hyperplane*. It does not have a physical representation in our 3D world. The way scientists present "*hyper-*"objects such as hyperplanes is by presenting consecutive 2D planes along different values of the 4th variable, the 5th variable, etc. This is why the use of functions, matrices and tensors is strictly needed to handle computations in multivariable spaces.

complicated ratio, fancy operator, square power, cubic, exponential, etc. This is what defines *linear* in mathematics –a.k.a. simple...

In our *3D* conception of the physical world the idea of linear only makes sense in *1D* (lines) and *2D* (planes). But for the purpose of computing, there is no need to "see" the model and any number of dimensions can be plugged-in, where one dimension represents one variable. For example, a hyperspace with 200 dimensions might be defined when looking at the results of a customer survey that contained 200 questions.

The regression method is thus a method of optimization which seeks the best approximation to a complex cloud of data points after reducing the complexity or number of equations used to describe the data. In high dimensions (i.e. when working with many variables), numerical optimization methods (e.g. Gradient Descent, Newton Methods) are used to find a solution by minimizing a so-called *loss function* [164]. But the idea is same as above: the loss function is either the Euclidean distance per se or a closely related function (developed to increase speed or accuracy of the numerical optimization algorithm [164]).

Complex relationship between dependent and independent variables can also be modeled via *non-linear* equations, but then the interpretability of the model becomes obscure because non-linear systems do not satisfy the *superposition* principle, that is the dependent variable is not directly proportional to the sum of the independent variable. This happens because at least one independent variable either appears several times in different terms of the regression equation or within some non-trivial operators such as ratios, powers, log. Often, alternative methods are preferable to a non-linear regression algorithm [165], see Chap. 7.

Complexity Tradeoff

The complexity of a model is defined by the nature of its representative equations, e.g. how many features are selected to make predictions and whether non-linear factors have been introduced. How complex the chosen model *should be* depends on a tradeoff between *under-fitting* (high bias) and *over-fitting* (high variance). In fact it is always theoretically possible to design a model that captures all idiosyncrasies of the cloud of data points, but such a model has no value because it does not extract any underlying trend [165], the concept of signal-to-noise ratio needs not apply... In contrast, if the model is too simple, information in the original dataset may be filtered out as *noise* when it actually represents relevant background signal, creating bias when making prediction.

The best regime between signal and noise often cannot be known in advance and depends on the context, so performance evaluation has to rely on a test-and-refine approach, for example using the method of *k*-folding described earlier in this chapter. Learning processes such as *k*-folding enable to seek the most robust model, i.e. the one that provides the highest score across the overall working set by taking advantage of as much information as possible in this set. They also enable to learn from (i.e. integrate) data acquired in real time.

6.2 Basic Probabilistic Tools and Concepts

Statistical inference [166] and probability theory [167] are the realm of forecasts and predictions. When a regression line is constructed and used to make predictions on the dependent variable (Eq. 6.7, or in the general case Eq. 6.9), the purpose shifts from simple description to *probabilistic inference*. A key underlying assumption in probability theory is that the dataset studied can be seen as a *sample* drawn from a larger dataset (in time and/or space), and can thus provide information on the larger dataset (in time and/or space).

The concept of p-value — or how to test hypotheses

One of the most common tools in data analysis is statistical *hypothesis testing* [166], which is an ubiquitous approach to deciding whether a given outcome is significant, and if yes with what level of confidence. The two associated concepts are *p-value* and *confidence interval*.

To first approximation, a statistical test starts with a hypothesis (e.g. *Juliet loves Romeo*), defines a relevant alternative hypothesis called the *null hypothesis* (e.g. *Juliet loves Paris* ...bad news for Romeo), and adopts a conservative approach to decide upon which is most likely to be true. It starts by assuming that the null hypothesis is true and that, under this assumption, a probability distribution exists for all variables considered (time spent between Romeo and Juliet, time spent between Paris and Juliet, etc), and this probability is not null. For example, if one variable is the time spent between Romeo and Juliet, it might be reasonable to assume that this quantity follows a normal distribution with mean of 2 h per week and standard deviation of 30 min, even after Paris proposed to Juliet. After all Romeo and Juliet met at the ball of the Capulet, they know each other, there is no reason to assume that they would never meet again. Then, it recons how likely it would be for a set of observations to occur if the null hypothesis was true. In our example, if after 3 weeks Romeo and Juliet spent 3 h together per week, how likely is it that Juliet loves Paris?

For a normal distribution, we know that about 68% of the sample lie within 1 standard deviation from the mean, about 95% lie within 2 standard deviations from the mean, and about 99% lie within 3 standard deviations from the mean. Thus in our example, the 95% confidence interval under the assumption that Juliet does love Paris (the null hypothesis) is between 1 h and 3 h. The probability to observe 3 h per week for three consecutive weeks in a row is $0.05 \times 0.05 \times 0.05 = 1.25 \times 10^{-4}$. *Thus there is < 0.1% chance to wrongly reject the null hypothesis.* From there, the test moves forward to conclude that Juliet *could* love Romeo because the null hypothesis has been rejected at the 0.001 level.

Reasonable threshold levels [168] for accepting/rejecting hypotheses are 0.1 (10%), 0.05 (5%), and 0.01 (1%). Statistical softwares compute p-values using a less comprehensive approach from the above, but more efficient when dealing with large datasets. Of course the theoretical concept is identical, so if you grasped it then whatever approach is used by your computer will not matter so much. Softwares rely on *tabulated ratios* that have been developed to fit different sampling distributions. For example, if only one variable is considered and a normal distribution with known standard deviation is given (as in the example above), a *z-test* is used, which relates the expected theoretical deviations (*standard error*[8]) to the observed deviations, rather than computing a probability for every observation as done above. If the standard deviation is not known, a *t-test* is adequate. Finally, if dealing with multivariate probability distributions containing both numerical and categorical (non quantitative, non ordered) variables, the generalized *χ-squared test* is the right choice. In data analytics packages, χ-square is thus often set as the default algorithm to compute p-values.

A critical point to remember about p-value is that it *does not prove a hypothesis* [169]: it indicates if an alternative hypothesis (called the null hypothesis, H_0) is more likely or not given the observed data and assumption made on probability distributions. That H is more likely than H_0 does not prove that H is true. More generally, a p-value is only as good as the hypothesis tested [168, 169]. Erroneous conclusions may be reached even though the p-values are excellent because of ill-posed hypotheses, inadequate statistics (i.e. assumed distribution functions), or sample bias.

Another critical point to remember about p-value is its *dependence on sampling size* [166]. In the example above, the p-value was conclusive because someone observed Romeo for 3 weeks. But on any single week, the p-value associated with the null hypothesis was 0.05, which would not be enough to reject the null hypothesis. A larger sample size *always* provides a lower p-value!

Statistical hypothesis testing, i.e. *inference*, should not be mistaken for the related concepts of *decision tree* (see Table 7.1) and *game theory*. The latters are also used to make decisions between events, but represent *less granular* methods as they themselves rely on hypothesis testing to assess the significance of their results. In fact for every type of predictive modeling (not only decision tree and game theory), p-values and confidence intervals are automatically generated by statistical softwares. For details and illustration, consult the application example of Sect. 6.3.

On Confidence Intervals — or How to Look Credible

Confidence intervals [91] are obtained by taking the mean plus and minus (right/left bound) some multiple of the standard deviation. For example in a normally distributed sample 95% of points lie within 1.96 standard deviations from the mean, which defines an interval with 95% confidence, as done in Eq. 7.17.

They provide a different type of information than p-value. Suppose that you have developed a great model to predict if an employee is a high-risk taker or is in contrast conservative in his decision makings (= response label of the model). Your model contains a dozen features, each with its own assigned weight, all of which have been selected with p-value <0.01 during the model design phase. Excellent. But your client informs you that it does not want to keep track of a dozen features on its employees, it just want about 2–3 features to focus on when meeting

[8]As mentioned in Sect. 6.1, the standard error is the standard deviation of the means of different sub-samples drawn from the original sample or population

prospective employees and quickly evaluate (say *without* a computer) their risk assertiveness. The confidence interval can help with this *feature selection*, because it provides information on the range of magnitude of the weight assigned to each feature. Indeed, if a confidence interval nears or literally includes the value *0*, then the excellent p-value essentially says that even though the feature is a predictor of the response variable, this feature is insignificant compared to some of the other features. The further away a weight is from *0*, the more useful its associated feature is for predicting the response label.

To summarize, the p-value does not provide any information whatsoever on how much each feature contributes, it just confirmed the hypothesis that these features have a positive impact on the desired prediction, or at least no detectable negative impact. In contrast— or rather to complement, confidence intervals enable the assessment of the magnitude with which each feature contributes relative to one another.

Central Limit Theorem — or Alice in Statistics-land

For a primer in data science, it is worth mentioning a widely applied theorem that is known as the foundation of applied statistics. The Central Limit Theorem [170] states that for almost every variable and every type of distribution, the distribution of the mean of these distributions will always be normally distributed for large sample size. This theorem is by far the most popular theorem in statistics (e.g. theory behind χ-squared tests and election polls) because many problems intractable for lack of knowledge on the underlying distribution of variables can be partially solved by asking alternative questions on the *means* of these variables [91]. Indeed the theorem says that the probability distribution for the mean of any variable is always perfectly known, it is a normal bell curve. Or almost always…

Bayesian Inference

Another key concept in statistics is the one of *conditional probability* and *Bayesian* modeling [167]. The intuitive notion of probability of an event is a simple *marginal* probability function [155]. But as for the concepts of marginal vs. conditional correlations described in Sect. 6.1, the behavior of a variable x might be influenced by its association with another variable y, and because y has its own probability function it does not have a static but a probabilistic influence on x. The actual probability of x knowing y is called the *partial probability of x given y*:

$$p(x|y) = \frac{p(y|x)\,p(x)}{p(y)} \tag{6.10}$$

where the difference between $p(x)$ and $p(x|y)$ is null only when the two variables are completely independent (i.e. their correlation $\rho(x,y)$ is null).

Without going into further details, it is useful to know that most predictive modeling softwares allow the user to introduce dependencies and prior knowledge (prior probability of event x) when developing a model. In doing so, you will be able to compute probabilities (posterior probability of event x) taking into account the effect of what you already know (for example if you know that a customer buys the Wall Street Journal every day, the probability that this customer buys The Economist too is not the world average, it is close to 1) and mutual dependencies (correlations) between variables.

6.3 Data Exploration

A data science project evolves through a standard number of phases [171], mainly *Collect, Explore, Clean, Process, Synthetize and Refine.*

The first three phases, *collect, explore, clean*, represent a phase of *data preparation* which enables and thus precedes the phases of processing and synthetizing insights from a dataset.

Collect

Data collection generally consists in *sampling* from a population. The actual population might be clearly defined, for example when polling for an electoral campaign, or be more conceptual, for example in weather forecast where the state in some location at time t is the sample and the state at that same location but at future time(s) is the population.

Statistical sampling is a scientific discipline in itself [172] and faces many challenges, some of which were discussed in Chap. 3, Sect. 3.2.3 that introduced *surveys*. For example, longitudinal and selection bias during data collection are commonly addressed by different types of random sampling techniques such as longitudinal stratified sampling and computer simulations [173]. Computer simulations will be introduced in the next chapter (Sect. 7.3).

Explore

The main objective of data exploration is to make choices. At the onset of any data science project indeed, choices need be made about the nature of each variable, and in particular one shall answer the following question: Is this variable potentially informative or not? If yes, what are potentially reasonable underlying probability distribution functions for this variable?

Data exploration consists in combining common sense with descriptive statistical tools. It includes visualization tools and basic theoretical underpinnings necessary to advance one's understanding of the data [91], such as recognizing peculiar distribution functions, defining outsiders and looking at cumulative functions. The reader may find it confusing at first because it overlaps with what one may expect to be a stage of *data processing*. And indeed data exploration is a never-ending process: the entire modeling apparatus enables information gathered at any stage of a data science project that may lead to valuable refinement to be fed back into the model.

But let us try to define the boundaries of this initial data exploration phase. Goals and bounds may be defined against the overall objectives of a data analysis project. Indeed the consultant, in order to solve a problem and answer questions addressed by the client, must choose a general category of modeling tools appropriate to the context and circumstances. This choice may be made through expert intuition or, conveniently for non-experts, the *Machine Learning* framework described in Chap. 7. This framework organizes, in a comprehensive way (see Table 7.1), some major classes of algorithms at disposal. Each algorithm comes with specific advantages, disadvantages, and a set of restraints and conditions. For example, most efficient predictive modeling algorithms only apply when the probability function underlying the distribution of input variables is normal (i.e. Gaussian [174]). From Table 7.1 thus, specific needs for data exploration emerge. For example one could consider that the phase of data exploration has been completed when one knows which approaches in Table 7.1 and Chap. 7 may be given a try!

Clean

Data cleaning addresses the issues of missing data, uninformative data, redundant data, too-noisy data and compatibility between different sources and formats of data.

The issue of *missing data* is typically addressed by deciding upon a satisfactory threshold for the number of observations available and applying this threshold to each variable considered. If the number of observations is below this threshold for any sample considered, the variable is discarded.

Non-informative variables can often be detected just by walking through the meaning of each variable in the dataset. For example, records of enrollment dates or "presence vs. absence" to a survey are often non-informative. The information contained in the questionnaire is useful but whether someone checked-in at this survey is likely not. This non-informative variable may thus be eliminated, which helps reduce model complexity and overall computing time.

The issue of *redundant* and *too-noisy* data can often be addressed by computing the marginal correlation between variables (Eq. 6.2). The goal is not to prove any relevant relationship but, in contrast, to filter out the non-relevant ones. For example, if two variables have a marginal correlation >0.95, the information provided by one of these variables shall be considered redundant. It will remain so for as long as the dataset does not change. If the project consists in developing a model based exclusively on the information contained in this dataset, Bayes rules (Eqs. 6.3 and 6.10) do not apply and thus nothing will change this overall marginal correlation of 0.95. The story is the same when a feature has a correlation <0.05 with the response label: at the end of the day nothing may change this overall characteristic of the data.

Thus, variables that are exceedingly redundant (too high ρ) or noisy (too low ρ) may be detected and eliminated at the onset of the project simply by looking at the marginal correlation ρ with the response label. Doing so saves a lot of time, it is akin to the 80/20 rule[9] in data science [175].

[9] The 80/20 rule, or Pareto principle, is a principle commonly used in business and economics that states that 80% of a problem stem from only 20% of its causes. It was first suggested by the late Joseph Juran, one of the most prominent management consultants of the twentieth century.

Last but not least, at the interface between data analysis and model design, the biggest challenges met by scientists handling large scale data (a.k.a. *big data*) today revolve less and less around data production and more and more around *data integration* [59, 65]. The key challenges are the integration of different formats, weights and meanings associated with different data sources relative to one another along the value networks. In this first quarter of the twenty-first century, there does not yet exist reliable standard or criteria in most circumstances to evaluate the importance of different *data sources* (i.e. to weigh different data sources) nor combine different *data formats* [176]. Most often it relies on intuition, which is not ideal given that most domain experts by definition specialize on some activities that produce specific data sources and never specialize in "all activities" at the origin of all these data sources [65].

Quantifying the utility of different data sources, and data integration in general, have become key objectives in most modern data science projects, far from just a preliminary step. Engineering data or metadata that enable better integration into a given model may itself rely on models and simulations. These methods are discussed in the next chapter.

Principles of Data Science: Advanced

7

This chapter covers advanced analytics principles and applications. Let us first back up on our objectives and progress so far. In Chap. 6, we defined the key concepts underlying the mathematical science of data analysis. The discussion was structured in two categories: descriptive and inferential statistics. In the context of a data science project, these two categories may be referred to as *unsupervised* and *supervised* modeling respectively. These two categories are ubiquitous because the objective of a data science project is always (bear with me please) to better understand some data or else to predict something. Chapter 7 thus again follows this binary structure, although some topics (e.g. computer simulation, Sect. 7.3) may be used to collect and understand data, forecast events, or both.

To better understand data, a data scientist may aim to transform how the data is described (i.e. filtering and noise reduction, Sect. 7.1) or re-organize it (clustering, Sect. 7.2). In both cases, data complexity can be reduced when signal vs. noise is detected. She/he may sample or forecast new data points (computer simulations and forecasting, Sect. 7.3). And more generally to predict events based on diverse, potentially very complex input data, she/he may apply statistical learning (machine learning and artificial intelligence, Sects. 7.4 and 7.5). A simple application of statistical learning for cost optimization in pharmaceutical R&D is given in Sect. 7.6 and a more advanced case on customer churn is given in Sect. 7.7.

The education system is so built that high-school mathematics tend to focus on key root concepts, with little to no application, and college-level mathematics tend to focus on applied methods. It often seems the why is obscure and the consequence pretty clear: most students don't like mathematics and quit before reaching a college-level education in mathematics, becoming forever skeptics about, well, what is the big deal with mathematics? In management consulting the 80/20 rule prevails so in this book the part on mathematics takes only two chapters (out of nine chapters, we practice what we preach). In Chap. 6 we covered the key root concepts. And in this chapter we focus on applied methods (a.k.a. college-level) including states of the art.

J. D. Curuksu, *Data Driven*, Management for Professionals,
https://doi.org/10.1007/978-3-319-70229-2_7

7.1 Signal Processing: Filtering and Noise Reduction

Signal processing means decomposing a signal into simpler components. Two categories of signal processing methods are in common usage, *Harmonic Analysis* and *Singular Value Decomposition*. They differ on the basis of their interpretability, that is, whether the building blocks of the original signal are known in advance. When a set of (simple) functions may be defined to decompose a signal into simpler components, this is Harmonic Analysis, e.g. *Fourier analysis*. When this is not possible and instead a set of generic (unknown) data-derived variables must be defined, this is Singular Value Decomposition, e.g. *Principal Component Analysis* (PCA). PCA is the most frequently used so let us discuss this method first.

Singular Value Decomposition (e.g. PCA)

An observation (a *state*) in a multivariable space may be seen as a point in a multidimensional coordinate system where each dimension corresponds to one variable. The values of each variable for a given observation are thereby the coordinates of this point along each of these dimensions. Linear algebra is the science that studies properties of coordinate systems by leveraging the convenient *matrix* notation, where a set of coordinates for a given point is called a *vector* and the multivariable space a *vector space*. A vector is noted $(x_1, x_2, \ldots, x_n)$ and contains as many entries as there are variables (i.e. dimensions) considered in the multivariable space.

A fundamental concept in linear algebra is the concept of *coordinate mapping* (also called isomorphism), i.e. the one-to-one linear transformation from one vector space onto another that permits to express a point in a different coordinate system without changing its geometric properties. To illustrate how useful coordinate mapping is in business analytics, consider a simple *2D* scatter plot where one dimension (x-axis) is the customer income level and the second dimension (y-axis) is the education level. Since these two variables are correlated, there exists a direction of maximum variance (which in this example is the direction at about 45° between the x- and y-axes because income- and education-levels are highly correlated). Therefore, rotating the coordinate system by something close to 45° will align the x-axis in the direction of maximum variance and the y-axis in the opposite, orthogonal direction of minimum variance. Doing so defines two new variables (i.e. dimensions), let us called them *typical* buyer profile (the higher the income, the higher the education) and *atypical* buyer profile (the higher the income, the lower the education). In this example the first new variable will deliver all information needed in this *2D* space while the second new variable shall be eliminated because its variance is much, much smaller. If your client is *BMW*, customer segments that might buy the new model are likely highly educated and financially comfortable, or poor and poorly educated, or in-between. But highly educated and financially poor customers are rare and unlikely to buy it, and even though poorly educated yet rich customers are certainly interested in the *BMW* brand, this market is even smaller. So the second variable may be eliminated because it brings no new information except for outsiders. By eliminating the second variable, we effectively reduced the number of dimension and thus simplified the prediction problem[1].

More generally, a set of observations in a multivariable space can always[2] be expressed in an alternative set of coordinates (i.e. variables) by the process of *Singular Value Decomposition*. A common application of SVD is the *Eigen-*decomposition[3] which, as in the example above, seeks the coordinate system along orthogonal (i.e. independent, uncorrelated) directions of maximum variance [177]. The new directions are referred to as eigenvectors and the magnitude of displacement along each *eigenvector* is referred to as *eigenvalue*. In other words, eigenvalues indicate the amount of dilation of the original observations along each independent direction.

$$Av = \lambda v \tag{7.1}$$

where $\boldsymbol{A}$ is a square $n \times n$ covariance matrix (i.e. the set of all covariances between n variables as obtained from Eq. 6.1), $\boldsymbol{v}$ is an unknown vector of dimension n, and $\boldsymbol{\lambda}$ is an unknown scalar. This equation, when satisfied, indicates that the transformation obtained by multiplying an arbitrary object by $\boldsymbol{A}$ is equivalent to a simple translation along a vector $\boldsymbol{v}$ of magnitude equal to $\boldsymbol{\lambda}$, and that there exists n pairs of $(\boldsymbol{v}, \boldsymbol{\lambda})$. This is useful because in this n-dimension space, the matrix $\boldsymbol{A}$ may contain non-zero values in many of its entries and thereby imply a complex transformation, which Eq. 7.1 just reduced to a set of n simple translations. These n vectors $\boldsymbol{v}$ are the characteristic vectors of the matrix $\boldsymbol{A}$ and thus referred to as its eigenvectors.

Once all n pairs of $(\boldsymbol{v}, \boldsymbol{\lambda})$ have been computed,[4] the highest eigenvalues indicate the most important eigenvectors (directions with highest variance), hence a quick look at the spectrum of all eigenvalues plotted in decreasing order of magnitude enables the data scientist to easily select a subset of directions (i.e. new variables) that have most impact in the dataset. Often the *eigen-spectrum* contains abrupt decays; these decays represent clear boundaries between more informative and less informative sets of variables. Leveraging the eigen-decomposition to create new variables and filter out less important variables reduces the number of variables and thus, once again, simplify the prediction problem.

---

[1] Note that in this example the two variables are so correlated that one could have ignored the other variable from the beginning and thereby bypass the process of coordinate mapping altogether. Coordinate mapping becomes useful when the trend is not just a 50/50 contribution of two variables (which corresponds to a 45° correlation line in the scatter plot) but some more subtle relationship where maximum variance lies along an asymmetrically weighted combination of the two variables.

[2] This assertion is only true under certain conditions, but for most real-world applications where observations are made across a finite set of variables in a population, these conditions are fulfilled.

[3] The word *Eigen* comes from the German for *characteristic*.

[4] The equation used to find eigenvectors and eigenvalues for a given matrix when they exist is $\det(A - \lambda I) = 0$. Not surprisingly, it is referred to as the matrix's *characteristic equation*.

The eigenvector-eigenvalue decomposition is commonly referred to as PCA (*Principal Component Analysis* [177]) and is available in most analytics software packages. PCA is widely used in signal processing, filtering, and noise reduction.

The major drawback of PCA concerns the interpretability of the results. The reason why I could name the new variables *typical* and *atypical* in the example above is that we expect income and education levels to be highly correlated. But in most projects, PCA is used to simplify a complex signal and the resulting eigenvectors (new variables) have no natural interpretation. By eliminating variables the overall complexity is reduced, but each new variable is now a composite variable born out of mixing the originals together. This does not pose any problem when the goal is to reconstruct a compound signal such as an oral speech recorded in a noisy conference room, because the nature of the different frequency waves in the original signal taken in isolation had no meaning to the audience in the first place. Only the original and reconstructed signals taken as *ensembles* of frequencies have meaning to the audience. But when the original components do have meanings (e.g. income levels, education levels), then the alternative dimensions defined by the PCA might loose interpretability, and at the very least demand new definitions before they may be interpreted.

Nevertheless, PCA analyses remain powerful in data science because they often entail a predictive modeling aspect which is akin to speech recognition in the noisy conference room: what matters is an efficient prediction of the overall response variable (the speech) rather than interpreting how a response variable relates to the original components.

Harmonic Analysis (e.g. FFT)

The SVD signal processing method (e.g. PCA) relies on a *coordinate mapping defined in vector space*. For this process to take place, a set of *data-derived* vectors (eigenvectors) and data-derived magnitudes of displacement (eigenvalues) need to be stored in the memory of the computer. This approach is truly generic in the sense that it may be applied in all types of circumstances, but becomes prohibitively computationally expensive when working with *very large* datasets. A second common class of signal processing methods, *Harmonic Analysis* [178], has smaller scope but is ultra-fast in comparison to PCA. Harmonic analysis (e.g. *Fourier analysis*) defines a set of *predefined* functions in the dataset that when superposed all together accurately re-construct or approximate the original signal. This technique works best when some localized features such as periodic signals can be detected at a macroscopic level[5] (this condition is detailed in the footnote).

[5]Quantum theory teaches us that everything in the universe is periodic! But describing the dynamics of any system except small molecules at a quantum level would require several years of computations even on last-generation supercomputers. And this is assuming we would know how to decompose the signal into a nearly exhaustive set of factors, which we generally don't. Hence an harmonic analysis in practice requires periodic features to be detected at a scale directly relevant to the analysis in question; this defines *macroscopic* in all circumstances. For example, a survey of customer *behaviors* may apply Fourier analysis if a periodic feature is detected in a behavior or any factor believed to influence a behavior.

In Harmonic analysis, an observation (a *state*) in a multivariable space is seen as the superposition of base functions called harmonic waves or *frequencies*. For example, the commonly used Fourier analysis [178] represents a signal by a sum of n trigonometric functions (sines and cosines), where n is the number of data points in the population. Each harmonic is defined by a frequency rate k and a magnitude a_k or b_k:

$$f(x) = a_0 + \int_{k=1}^{n} \left(a_k \cos\left(kc_0\pi x\right) + b_k \sin\left(kc_0\pi x\right) \right) \tag{7.2}$$

The coefficients of the harmonic components (a_k, b_k) can easily be stored which significantly reduce the total amount of storage/computational power required to code a signal compared to PCA where every component is coded by a pair of eigenvector and eigenvalue. Moreover, the signal components (i.e. harmonic waves) are easy to interpret, being homologous to the familiar notion of frequencies that compose a music partition (this is literally what they are when processing audio signals).

Several families of functions that map the original signal into the *frequency domain*, referred to as *transforms*, have been developed to fit different types of application. The most commons are Fourier Transform (Eq. 7.2), FFT (Fast Fourier Transform), Laplace Transform and Wavelet Transform [179].

This main drawback of Harmonic Analysis compared to PCA is that the components are not directly derived from the data. Instead, they rely on a predefined model which is the chosen Transform formula, and thus may only reasonably re-construct or approximate the original signal under the presence of macroscopically detectable periodic features (see footnote on previous page; these signals are referred to as *smooth* signals).

Re-constructing an original signal by summing up all its individual components is referred to as a *synthesis* [178], by opposition to an *analysis* (a.k.a. deconstruction of the signal). Note that Eq. 7.2 is a synthesis equation because of the integral in front, i.e. the equation used when reconstructing the signal. Synthesis may be leveraged in the same way as PCA by integrating only the *high-amplitude* frequencies and filtering out the *low-amplitude* frequencies, which reduces the number of variables and thus simplify the prediction problem.

As for PCA, harmonic analysis and in particular FFT is available in most analytics software packages, and is a widely used technique for signal processing, filtering and noise reduction.

7.2 Clustering

The process of finding patterns and hidden structures in a dataset is often referred to as clustering, partitioning, or *unsupervised* machine learning (by opposition to *supervised* machine learning described in Sect. 7.4). Clustering a dataset consists in grouping the data points into subsets according to a distance metric such that data points in the same subset (referred to as *cluster*) are more similar to each other than to points in other clusters.

A commonly used *metric* to define clusters is the Euclidean distance defined in Eq. 6.6, but there are in fact as many clustering options as there are available metrics.

Different types of clustering *algorithms* are in common usage [180]. Two common algorithms, *k*-mean and hierarchical clustering, are described below. Unfortunately, no algorithm completely solves for the main drawback of clustering which is to choose the best number of clusters for the situation at hand. In practice, the number of clusters is often a fixed number chosen in advance. In hierarchical clustering the number of clusters may be optimized by the algorithm based on a threshold in the value of the metric used to define clusters, but since the user must choose this threshold in advance this is just a chicken and egg distinction. No one has yet come up with a universal standard for deciding upon the best number of clusters [180].

In *k-mean* (also referred to as *partitional clustering* [180]), observations are partitioned into *k* clusters[6] by evaluating the distance metric of each data point to the mean of the points already in the clusters. The algorithm starts with dummy values for the k cluster-means. The mean that characterizes each cluster, referred to as *centroid*, evolves as the algorithm progresses by adding points in the clusters one after the other.

For very large datasets, numerical optimization methods may be used in order to find an optimum partitioning. In these cases, the initial dummy value assigned to the k starting centroids should be chosen as accurately as intuition or background information permits in order for the k-mean algorithm to converge quickly to a local optimum.

In *hierarchical clustering* [180], the observations are partitioned into k clusters by evaluating a measure of connectivity (a.k.a. dissimilarity) for each data point between the clusters. This measure consists in a distance metric (e.g. Euclidean) and a *linkage criteria* (e.g. average distance between two clusters). Once the distance and linkage criteria have been chosen, a *dendrogram* is built either top down (*divisive* algorithm) or bottom up (*agglomerative* algorithm). In the top down approach, all observations start in one cluster and splits are performed recursively as one moves down the hierarchy. In the bottom up approach in contrast, each observation starts in its own cluster and pairs of clusters merge as one moves up the hierarchy.

In contrast to k-mean, the number of clusters in a hierarchical clustering needs not be chosen in advance as it can more naturally be optimized according to a threshold for the measure of connectivity. If the user wishes so, he/she may however choose a pre-defined number of clusters in which case no threshold is needed; the algorithm will just stop when the number of clusters reaches the desired target.

One advantage of hierarchical clustering compared to k-mean is the interpretability of results: when looking at the hierarchical dendrogram, the relative position of every cluster with respect to one another is clearly presented within a

[6] *k* can be fixed in advance or refined recursively based on a distance metric threshold.

comprehensive framework. In k-mean in contrast, the closeness of the different clusters with respect to one another may be impossible to articulate if there are more than just a few clusters.

Most analytics software packages offer *k*-mean and hierarchical clustering platforms. Hierarchical clustering offers better flexibility in term of partitioning options (choice between distance metrics, linkage criteria and top down vs. bottom up) and better interpretability with respect to k-mean clustering, as explained above. But both remain widely used [180] because the complexity of hierarchical search algorithms makes them too slow for large datasets. In this case, a potential tactic may be to start with k-mean clustering, sample randomly within each of the k clusters, and then apply a hierarchical search algorithm. Again, it is not about academic science, it is about management consulting. The 80/20 rule prevails.

7.3 Computer Simulations and Forecasting

Forecasts may be carried out using different methods depending on how much detail we know on the probability distribution of the data we aim to forecast. This data is often a set of quantities or coordinates characterizing some event or physical object, and is thus conveniently referred to as the past, present and future *states* of a given system. If we had a perfectly known probability density function for all components of the system, for a given initial state all solutions at all times (called closed form solutions) could be found. Of course this function we never have. So we use numeric approximation, by discretizing space and time into small intervals and computing the evolution of states one after the other based on available information. Information generally used to predict the future includes *trends* within the past evolution of states, *randomness* (i.e. variability around trends) and *boundary conditions* (e.g. destination of an airplane, origin of an epidemic, strike price of an option, low energy state of a molecule, etc). Auto-regressive models (7.3.1) can predict short sequences of states in the future based on observed trends and randomness in the past. Finite difference methods (7.3.2) can create paths of states based on boundary conditions by assuming Markov property (i.e. state at time t only depends on state at previous time step), or more detailed trajectories by combining boundary conditions with some function we believe to approximate the probability distribution of states. Monte Carlo sampling (7.3.3) in contrast may not reconstruct the detailed evolution of states, but can efficiently forecast expected values in the far future based on simple moments (mean, variance) of the distribution of states in the past, together with some function we believe drive the time evolution of states. Such function is referred to as a *stochastic process*. This is a fundamental building block in many disciplines such as mathematical finance: since we can never know the actual evolution of states [of a stock], the process should include a *drift* term that drives what we know on deterministic trends and a *random* term that accounts for multiple random factors that we cannot anticipate.

7.3.1 Time Series Forecasts

When time series in the past is available and we want to extrapolate the time series in the future, a standard method consists in applying regression concepts from the past states onto the future states of a variable, which is called auto-regression (AR). This auto-regression can be defined on any p number of time steps in the past (p-th order Markov assumption, i.e. only p lags matter) to predict a sequence of n states in the future, and is thus deterministic. Given we know and expect fluctuation around this mean predication due to multiple random factors, a stochastic term is added to account for these fluctuations. This stochastic term is usually a simple random number taken from a standard normal distribution (zero mean, unit standard deviation), called *white noise*.

Since the difference between what the deterministic, auto-regressive term predicts and what is actually observed is also a stochastic process, the auto-regression concept can be applied to predict future fluctuations around the predicted mean based on past random fluctuations around the past means. In other words, to predict future volatility based on past volatility. This makes the overall prediction capture uncertainties not just at time t but also on any q number of time steps in the past (q-th order Markov assumption). This term is called moving average (MA) because it accounts for the fact that the deterministic prediction of the mean based on past data is a biased predictor: in fact, the position of the mean fluctuates within some evolving range as time passes by. MA adjusts for this stochastic evolution of the mean.

Finally, if an overall, non-seasonal trend (e.g. linear increase, quadratic increase) exists, the mean itself evolves in time which may perpetually inflate or deflate the auto-regressive weights applied on past states (AR) and fluctuations around them (MA). So a third term can be added that takes the difference between adjacent values (corresponds to first order derivative) if the overall trend is linear, the difference between these differences (second order derivative) if the overall trend is quadratic, etc. The time series is integrated (I) in this way so that AR and MA can now make inference on time series with stable mean. This defines ARIMA [181]:

$$x(t) = \int^{p} a_i x_{t-i} + \varepsilon_t + \int^{q} b_j \varepsilon_{t-j} \tag{7.3}$$

where the first integral (i.e. sum) is the deterministic part AR and the other integral is the stochastic part MA, p and q are the memory spans (a.k.a. lags) for AR and MA respectively, a_i are the auto-regression coefficients, ε_i are white noise terms (i.e. random samples from normal distributions with mean of 0 and standard deviation of 1) and x_t is the observed, d-differenced stationary stochastic process. There exist many variants of ARIMA such as ARIMAx [182] ('x' stands for exogenous inputs) where auto-regression Eq. 7.3 is applied both on the past of the variable we want to predict (variable x) and on the past of some other variables (variables z_1, z_2, etc) that we believe to influence x; or SARIMA where Eq. 7.3 is modified to account for seasonality [183].

The main limits of ARIMA approaches are the dependence on *stationary* data (mean and probability distribution is invariant to shifting in time) and on *mixing* (correlation between states vanishes after many time steps so that two such states become independent events) [184]. Indeed the simple differencing explained above does not guarantee stationary data. In fact, it is almost never the case that a time series on average increases or decreases exactly linearly (or quadratically, etc). So when the differencing in ARIMA is carried out, there is always some level of non-stationarity left over. Moreover, if the time series is complex or the dependence between variables in ARIMAX is complex, a simple auto-regression approach will fail. Then some non-parametric time series forecasting methods have a better chance to perform well, even though they don't offer a clear interpretable mapping function between inputs and outputs as in Eq. 7.3. We present a new generation non-parametric approach for time series forecasting (recurrent deep learning) in Sect. 7.4.

7.3.2 Finite Difference Simulations

Finite difference methods simulate *paths* of states by iteratively solving the derivative of some function that we believe dictate the probability distribution of states across space S and time t:

$$\frac{\partial f}{\partial S} = \frac{f_{i,j+1} - f_{i,j}}{\Delta S}, \frac{\partial f}{\partial t} = \frac{f_{i+1,j} - f_{i,j}}{\Delta t}, \frac{\partial^2 f}{\partial S^2} = \frac{f_{i,j+1} - f_{i,j-1} - 2f_{i.j}}{\Delta S^2} \tag{7.4}$$

where a dynamic state (i, j) is defined by time $t = i$ and space $S = j$. Let us look at a few examples to see how Eq. 7.4 can be used in practice. A simple and frequent case is the absence of any information on what a function f could be. One can then assume stochastic fluctuations to be uniform in space and time except for small non-uniform difference in space observed at instantaneous instant t. This non-uniformity, in the absence of any additional net force acting upon the system, will tend to diffuse away, leading to gradual mixing of states (i.e. states become uncorrelated over long time period) called *dynamic equilibrium* [184]. It is common to think about this diffusion process as the diffusion of molecules in space over time [185], with the temperature acting as the uniform stochastic force that leads molecules to flow from regions of high concentration toward regions of low concentration until no such gradient of concentration remain (*thermal equilibrium*). In a diffusion process, fluctuations over time are related to fluctuations over space (or stock value, or any other measure) through the *Diffusion equation*:

$$\frac{\partial f}{\partial t} = D(t)\nabla^2 f \tag{7.5}$$

In many cases $D(t)$ is considered constant, which is the *Heat equation*. By combining Eqs. 7.4 and 7.5, we can express f at time $t + 1$ from f at time t even if we don't know anything about f: all we need are some values of f at some given time t and the boundary conditions. The "non-uniform difference" in space observed at time t will be used to compute the value at a time $t + 1$, and all value until any time in the future, one step at a time:

$$f_j^{i+1} = \alpha f_{j-1}^i + (1-2\alpha) f_j^i + \alpha f_{j+1}^i \tag{7.6}$$

where $\alpha = \Delta t/(\Delta x)^2$. Similar to Eq. 7.6, we can write a backward finite difference equation if boundary conditions are so given that we don't know the initial values but instead we know the final values (e.g. strike price of an option), and create paths going backward in time.

Now, if we do have an expression for f that we believe approximate the probability distribution of states, we can use a Taylor expansion to equate the value of f with its first order derivatives[7] and use Eq. 7.4 to equate the first order derivatives of f with the state at time t and $t + 1$.

$$f(x) = f(x_0) + f'(x_0)(x - x_0) + \frac{f''(x_0)}{2!}(x - x_0)^2 + \ldots + \frac{f^{(n)}(x_0)}{n!}(x - x_0)^n \tag{7.7}$$

A popular example is the Newton's method used to find the roots of f:

$$x_{i+1} = x_i - \frac{f(x_i)}{f'(x_i)} \tag{7.8}$$

Equation 7.8 is an iterative formula to find x for which $f(x) = 0$, and derived by truncating the Taylor series at its first order and taking $f(x) = 0$, which expresses any x as function of any x_0, taken respectively to be x_{i+1} and x_i.

Let us look at two concrete examples. First at a specific example in Finance, delta-hedging, a simple version of which consists in calling an option and simultaneously selling the underlying stock (or vice-versa) by a specific amount to hedge against volatility, leading to an equation very similar to Eq. 7.5 except that *there are* external forces acting upon the system. An option can be priced by assuming no-arbitrage [185], i.e. a theoretical "perfect" hedging: the exact quantity of stock to sell to hedge against the risk of losing money with the given option is being sold at all time. This quantity depends on the current volatility, which can never be known perfectly, and needs be adjusted constantly. Hence, arbitrage opportunities always exist in reality (it is what the hedge fund industry is built upon). The theoretical price of an option can be based on no-arbitrage as a reference, and this leads (for a demonstration see Refs. [185, 186]) to the following *Black Scholes* formula for the evolution of the option price:

$$\frac{\partial f}{\partial t} + rS\frac{\partial f}{\partial S} + \frac{1}{2}\sigma^2 S^2 \frac{\partial^2 f}{\partial S^2} = rf \tag{7.9}$$

Equation 7.9 mainly differs from Eq. 7.5 by additional terms weighted by the volatility of the underlying stock (standard deviation σ) and the risk-free rate r. Intuitively, think about r as a key factor affecting option prices because the volatility

[7] It is standard practice in calculus to truncate a Taylor expansion after second order derivative because higher order term tend to be insignificant.

of the stock is hedged, so the risk free rate is what the price of an option depends on under the assumption of no-arbitrage. We may as before replace all derivatives in Eq. 7.9 by their finite difference approximation (Eq. 7.4), re-arrange the terms of Eq. 7.9, and compute the value at time t + 1, and all value until any time in the future, one step at a time:

$$f_j^{i+1} = a_j f_{j-1}^i + b_j f_j^i + c_j f_{j+1}^i \tag{7.10}$$

where a_j, b_j, c_j are just the expressions obtained when moving all but the $i + 1$ term on the right hand side of Eq. 7.9.

Finally, let us now look at a more general example, that may apply as much in chemistry as in finance, a high-dimensional system (i.e. a system defined by many variables) evolving in time. If we know the mean and covariance (or correlation and standard deviation given Eq. 6.2) for each component (e.g. each stock's past mean, standard deviation and correlation with each other), we can define a function to relate the probability of a given state in the multivariable space to a *statistical density potential* that governs the relationship between all variables considered [187]. This function can be expressed as a simple sum of harmonic terms for each variable as in Eq. 7.11 [188], assuming a simple relationship between *normally distributed* variables:

$$\begin{aligned} E(x_1, x_2, \dots, x_n) = & \int_{x_1} \operatorname{cov}(x_{i1})^{-1} (x_{i1} - \bar{x}_1)^2 \\ & + \int_{x_2} \operatorname{cov}(x_{i2})^{-1} (x_{i2} - \bar{x}_2)^2 + \dots \\ & + \int_{x_n} \operatorname{cov}(x_{in})^{-1} (x_{in} - \bar{x}_n)^2 \end{aligned} \tag{7.11}$$

If we think about this density potential as the "energy" function of a physical system, we know that high-energy *unstable* states are exponentially unlikely and low-energy *stable* states are exponentially more likely (this is a consequence of the *canonical* Boltzmann distribution law[8] [188]). In theoretical physics, the concepts of *density estimation* and *statistical mechanics* provide useful relationship between microscopic and macroscopic properties of high dimensional systems [187], such as the probability of the system:

$$p(x_n) = \frac{e^{\frac{-E}{k_B T}}}{\int e^{\frac{-E}{k_B T}}} \tag{7.12}$$

[8] The idea that stable states are *exponentially* more likely than unstable states extends much beyond the confines of physical systems. This *Boltzmann distribution law* has a different name in different fields, such as the Gibbs Measure, the Log-linear response or the Exponential response (to name a few), but the concept is always the same: *There is an exponential relationship between the notions of probability and stability.*

where E is the density potential (energy function) chosen to represent the entire system. The integral in the denominator of Eq. 7.12 is a normalization factor, a sum over all states referred to as the *partition function*. The partition function is as large as the total number of possible combinations between all variables considered, and thus Eq. 7.12 hold only as much as dynamic equilibrium is achieved, meaning the sample generated by the simulation should be large enough to include all low energy states because these contribute the most to Eq. 7.12. Dynamic equilibrium, or ergodicity, is indeed just a cool way to say that our time series owes to be a representative sample of a general population, with all important events sampled.

To generate the sample, we can re-write Eq. 7.7 in terms familiar to physics:

$$x(t+\delta t) = x(t) + v(t)\delta t + \frac{1}{2}a(t)\delta t^2 + O(\delta t^3) \tag{7.13}$$

Equation 7.13 expresses the coordinates of a point in a multidimensional system (i.e. an observed state in a multivariable space) at time $t + 1$ from its coordinates and first-order derivatives at time t [189], where $v(t)$ represents a *random perturbation* (stochastic frictions and collisions) that account for context-dependent *noise* (e.g. overall stock fluctuations, temperature fluctuations, i.e. any random factor that is not supposed to change abruptly), and $a(t)$ represents the *forces* acting upon the system through Newton's second law:

$$F(t) = -\nabla E(\mathrm{t}) = m \times a(t) \tag{7.14}$$

The rate of change of the multivariable state x and the evolution of this rate can be quantified by the first- and second-order derivatives of x, respectively $v(t)$ and $a(t)$ in Eq. 7.13. In large datasets, a set of initial dummy *velocities* $v_0(t)$ may be assigned to start the simulation as parts of the boundary conditions and updated at each time step through finite difference approximation. The derivative of the density potential $E(t)$ defines a "force field" applied on the coordinates and velocities, i.e. the *accelerations* $a(t)$, following Eq. 7.14.

As in all other examples discussed in this section, Eq. 7.13 computes the value at time $t + 1$, and all values until any time in the future, one step at a time. The dynamics of the system is numerically simulated for a large number of steps in order to equilibrate and converge the rates of change $v(t)$ and $a(t)$ to some stationary values [189]. After this equilibration phase, local optima can be searched in the hyperspace, random samples can be generated, and predictions of future states may be considered.

The result of numeric computer simulation techniques in multivariable environment thus consists in a random walk along the hyperspace defined by all the variables [188, 189]. The size of the time-step is defined by the level of time-resolution, which means the fastest motion in the set of dynamically changing variables $(x_1, x_2, \ldots, x_n)$ explicitly included in the density potential E [189]. In the rare cases where the density potential is simple enough for both first-order derivatives (*gradient* matrix) and second order-derivatives (*hessian* matrix) to be computed, deterministic simulations (i.e. methods that exhaustively run through the entire population) may be used, such as *Normal Mode* [190]. But numeric methods are by far more common.

Most analytics software packages include algorithms to carry out simulations of multivariable systems and produce random samples. They offer many options to customize the evolution equations in the form of ordinary and partial differential equations (ODE, PDE). Optimization algorithms, e.g. Stochastic Gradient Descent and Newton methods, are also readily available in these packages.

The main drawback of finite difference simulations, both for optimization, prediction and random sampling, revolves around the accuracy of the density potential chosen to represent the multivariable system [189], or the evolution equations (i.e. the Diffusion and Black Scholes equations in first two examples). Rarely are all intricacies of the relationships between all variables or random factors captured, and when the definition of the system attempts to do so the equations involved become prohibitively time consuming. As an alternative to dynamic finite difference simulation, Monte Carlo can be used. Monte Carlo will not enable the analysis of individual trajectories. But if what really matters is the expected value of some statistics over an ensembles of representative trajectories, then Monte Carlo is likely the best option.

7.3.3 Monte Carlo Sampling

The Monte Carlo method is widely used to generate samples that follow a particular probability distribution [185]. The essential difference with the finite difference method is that the detailed time-dependent path does not need to be (and is generally not) followed precisely, but can be evaluated using random numbers with precise probability. It is interesting to see how very simple expressions for this probability can in practice solve formidably complex deterministic problems. This is made possible by relying on the so-called law of large numbers which postulates that values sampled through a large number of trials converge toward their expected values, regardless of what factors influence their detailed time evolution [185]. Of course if the detailed dynamic is required, or events/decisions need be modeled along the trajectory, Monte Carlo is not the method of choice. But to evaluate expected values of certain processes, it opens the door to highly simplified, highly efficient solutions.

Let us look at two concrete examples, and close with a review of main pros and cons with Monte Carlo. The first example is very common in all introductions to Monte Carlo: approximate the value of pi. The algorithm essentially relies on two ingredients: a *stochastic process*, a repeated trial of a number that follows a uniform distribution between 0 and 1, and one simple *formula for the expected value:* Area of a circle = $r^2 \times$ pi. Take the radius to be 1 and imagine a square of length 2 circumscribing the circle. By placing the center of the circle at the origin *(0, 0),* and defining random numbers sampled between 0 and 1 as the *(x, y)* coordinates of some points in or out of the circle, the ratio of the circle's area on the square's area = pi/4. This ratio can be easily counted (all points inside the circle have norm ≤ 1). Once we have sampled few thousand points, the average of this ratio becomes quite accurate (i.e. by the law of large numbers), and since pi equals four times this ratio, so does our estimate of pi.

As second example, let's again look into finance and how Monte Carlo can be used to price an option. In this case, instead of a uniform probability between 0 and 1, the stochastic process contains much more information, mainly a time component that depends on both the stock price and its average value (until present day), and a space component that depends on the stock price and it standard deviation (i.e. volatility):

$$dS = \hat{\mu} S dt + \sigma S dz \tag{7.15}$$

$$S(t+\Delta t) - S(t) = \hat{\mu} S(t)\Delta t + \sigma S(t)\varepsilon\sqrt{\Delta t} \tag{7.16}$$

where $\hat{\mu}$ is the expected return and σ is the volatility. Using the equivalence noted earlier between space and time components in diffusion processes, which most of finance theory follows, Eq. 7.15 can be discretized in to Eq. 7.16, where ε is white noise, i.e. a random sample from a normal distribution with mean of 0 and standard deviation of 1.

Equation 7.16 enables to compute the value of the stock at time $t + 1$, which in turn enable to compute it at $t + 2$, and so forth. A Monte Carlo simulation 'trial' here involves constructing a complete path for the stock value using n random samples from a normal distribution. Once this is done, the payoff from the option can be calculated at a later time T, the maturity time of the option. And once we have sampled a few thousand paths/payoffs, the average of this payoff becomes quite accurate (i.e. by the law of large numbers) [185, 186]. So does our estimate of the option value, which is the present value of the expected payoff obtained using risk free rate as it assumes a proper hedging against volatility, see details in previous section.

Monte Carlo has many advantages compared to finite difference and other forecasting methods. First, because it is based on the law of large number it can handle complex systems with much simpler descriptions (e.g. uniform and stochastic process respectively for examples above). It is numerically efficient, in particular for processes with multiple variables. Monte Carlo can be combined with SVD to easily create samples of correlated variables [185, 186], which is very useful in Finance to model *portfolios* of securities. The running time increases linearly with the number of variables, while for most other procedures the running time increases exponentially with the number of variables.

Second, the central limit theorem (see *Alice in statistics land* in previous chapter) applies fully in Monte Carlo since it consists in simulating a large number of samples and taking the average of converged quantities across these samples. By the central limit theorem, these quantities follow a normal distribution regardless of their original generation process, and thereby the standard error can be used to provide a confidence interval (see *How to look credible* in previous chapter). In the example of predicting option prices, a 95% confidence interval for the price f is given by:

$$\mu - \frac{1,96\sigma}{\sqrt{n}} < f <= \mu + \frac{1,96\sigma}{\sqrt{n}} \tag{7.17}$$

where μ is the present value of the expected payoff, σ is the standard deviation, and n is the number of Monte Carlo samples.

Another strength of Monte Carlo is that it can be used to estimate some function of the whole path followed in each sample, not only their terminal value [186].

There are essentially two main drawbacks of Monte Carlo. First, it can still be very time consuming even though faster than more explicit, or deterministic, procedures. In Eq. 7.17, we see that to increase the accuracy by a factor of 10, the number of samples must increase by a factor of 100, and so on. Second, Monte Carlo provides only stochastic paths rationalized by probability concepts (Markov property), not actual sequences of simulated events. To sample and simulate sequential or multi-stage decision problems, finite difference and other dynamic programming approaches might be preferable.

Sampling Techniques to Improve Performance

To close this discussion on Monte Carlo and simulation techniques, let us briefly mention some standard procedures to reduce the number of samples needed in Monte Carlo, and more generally to increase convergence in computer simulations. *Stratified and quasi-random sampling* are based on sampling "representative" values from sub-samples distributed uniformly in the probability space, rather than sampling this probability space with truly random trials. *Importance sampling* restrains the sampling to some region of the probability space known to be more realistic or desired, then multiply expected values by the probability of these restraints to be observed following Bayesian rules (Eq. 6.10 in previous chapter). The *moment matching* techniques [191] consist in adjusting the samples to match the mean and standard deviation (and higher order moments) of some reference probability distribution. The *control variate* procedure [192] consists in simulating both the quantity of interest and another 'reference' quantity which is known to be similar and for which we have a deterministic (or somehow more reliable) solution. Both simulations use the exact same random number at every step, and the quantity of interest is adjusted by adding the difference between true vs. simulated reference quantity. Accuracy is improved because if the two quantities are indeed similar, they should experience similar bias at each step. A similar concept is used in the *antithetic variable* approach [193], where at every step the quantity of interest is simulated twice by taking the actual and opposite sign of the random number, and then taking the average between the two. It is expected that with the law of large number, when one tends to be higher than the true value, the other tends to be lower, so averaging the two increases accuracy after many time steps.

Finally, note that if the above methods aim at reducing variance, some problems may require just the opposite. Increasing variance may be needed when there are so many variables that there exists more than one expected value or high probability "level state". This is called a multi minima problem. Let's introduce *replica exchange* [192, 194] as one example: here a finite difference sampling is combined with stochastic swaps of states between different replicas of the simulation, each running with different parameters (e.g. same probability function but higher variance, shifted mean, etc). Each stochastic swap aims at providing new starting points in the replica of interest [194], so that this replica gets opportunities to explore more diverse regions of the complex multi-dimensional probability space, and in so doing identify multiple outcomes with high probability [192, 194].

7.4 Machine Learning and Artificial Intelligence

Machine learning is currently the most popular buzzword in data science, or the *science of predictions* [194]. *Supervised* machine learning is often used to mean all of the predictive modeling arsenal [194], by opposition to the more descriptive phases of data exploration (Sect. 6.3). Most descriptive methods are themselves referred to as *unsupervised* machine learning. So, pretty much all of the mathematics can be framed under the effigy of machine learning [195].

Most tools upon which machine learning is founded were already known long before the term machine learning was coined for the first time. The reader might have never heard about machine learning, yet he or she will probably have heard about the tools: regression, hypothesis tests, clustering, bootstrapping, neural networks, Bayesian inference, decision trees. The strength of the machine learning framework is to bring it all together in one comprehensive, all-encompassing structure. This discipline is born out of recent developments in the field of robotics with new algorithms behind pattern recognition and experience-based learning that underlie the decision-making process of intelligent machines [195]. It is not surprising that researchers in robotics would seek to develop a general framework aimed at taking advantage of all the mathematical arsenal at disposal since it has now unlocked a series of new generation and ubiquitous smart devices.

7.4.1 Overview of Models and Algorithms

Unsupervised Methods (Clustering and Spectral Analysis)

Unsupervised machine learning [195] includes all the techniques of clustering, filtering, and noise reduction introduced in the previous section (e.g. *k-means*, *PCA*, *FFT*). These methods have in common that there is no *response variable* (*label*) to lead and evaluate performance of the model [195]. The objective of unsupervised problems is to find hidden patterns and signals in a complex dataset. For example one is looking for high density populations (clusters), high variance dimensions (PCA), or dominant harmonics (Fourier).

Supervised Methods

For a model to qualify as a supervised machine learning algorithm, it has to *learn* from the processed data to make predictions [194], by opposition to a model parameterized by instructions predefined without influence from the data. Of course, all models are always defined under the influence of "some" data (the intuition of the programmer at least!), but to qualify the algorithm needs to explicitly take into account a *training dataset* when developing the *inference function*[9], i.e. the equation that will map some input variables, the *features*, into some output variable(s), the *response label*. The method of *k-folding* may be used to define a training set and a testing set, see Sects. 6.1 and 7.4.2 for details. The inference function is the actual problem that the algorithm is trying to solve [195] – for this reason this function is referred to as the *hypothesis h* of the model:

$$h: (x_1, x_2, \ldots, x_{n-1}) \mapsto h(x_1, x_2, \ldots, x_{n-1}) \sim x_n \qquad (7.18)$$

where $(x_1, x_2, \ldots, x_{n-1})$ is the set of features and $h(x_1, x_2, \ldots, x_{n-1})$ is the predicted value for the response variable(s) x_n. The hypothesis h is the function used to make predictions for as-yet-unseen situations. New situations (e.g. data acquired in real-time) may be regularly integrated within the training set, which is how a robot may learn in real time – and thereby *remember*... (Fig. 7.1).

Regression vs. Classification

Two categories of predictive models may be considered depending on the nature of the response variable: either *numerical* (i.e. response is a number) or *categorical* (i.e. response is not a number[10]). When the response is numerical, the model is a regression problem (Eq. 6.9). When the response is categorical, the model is a classification problem. When in doubt (when the response may take just a few ordered labels, e.g. 1, 2, 3, 4), it is recommended to choose a regression approach [165] because it is easier to interpret.

In practice, choosing an appropriate framework is not as difficult as it seems because some frameworks have clear advantages and limitations (Table 7.1), and because it is always worth trying more than one framework to evaluate the robustness of predictions, compare performances and eventually build a compound model made of best performers. This approach of blending several models together is itself a sub-branch of machine learning referred to as *Ensemble* learning [165].

Selecting the Algorithm

The advantages and limitations of machine learning techniques in common usage are discussed in this section. Choosing a modeling algorithm is generally based on four criteria: accuracy, speed, memory usage and interpretability [165]. Before considering these criteria however, considerations need be given to the nature of the variables. As indicated earlier, a regression or classification algorithm will be used depending on the nature (numerical or categorical) of the response label. Whether

[9] The general concept of inference was introduced in Sect. 6.1.2 – Long story short: it indicates a transition from a *descriptive* to a *probabilistic* point of view.

[10] Categorical variables also include *non-ordinal* numbers, i.e. numbers that don't follow a special order and instead correspond to different *labels*. For example, to predict whether customers will choose a product identified as *#5* or one identified as *#16*, the response is presented by two numbers (5 and 16) but they define a *qualitative* variable since there is no unit of measure nor zero value for this variable.

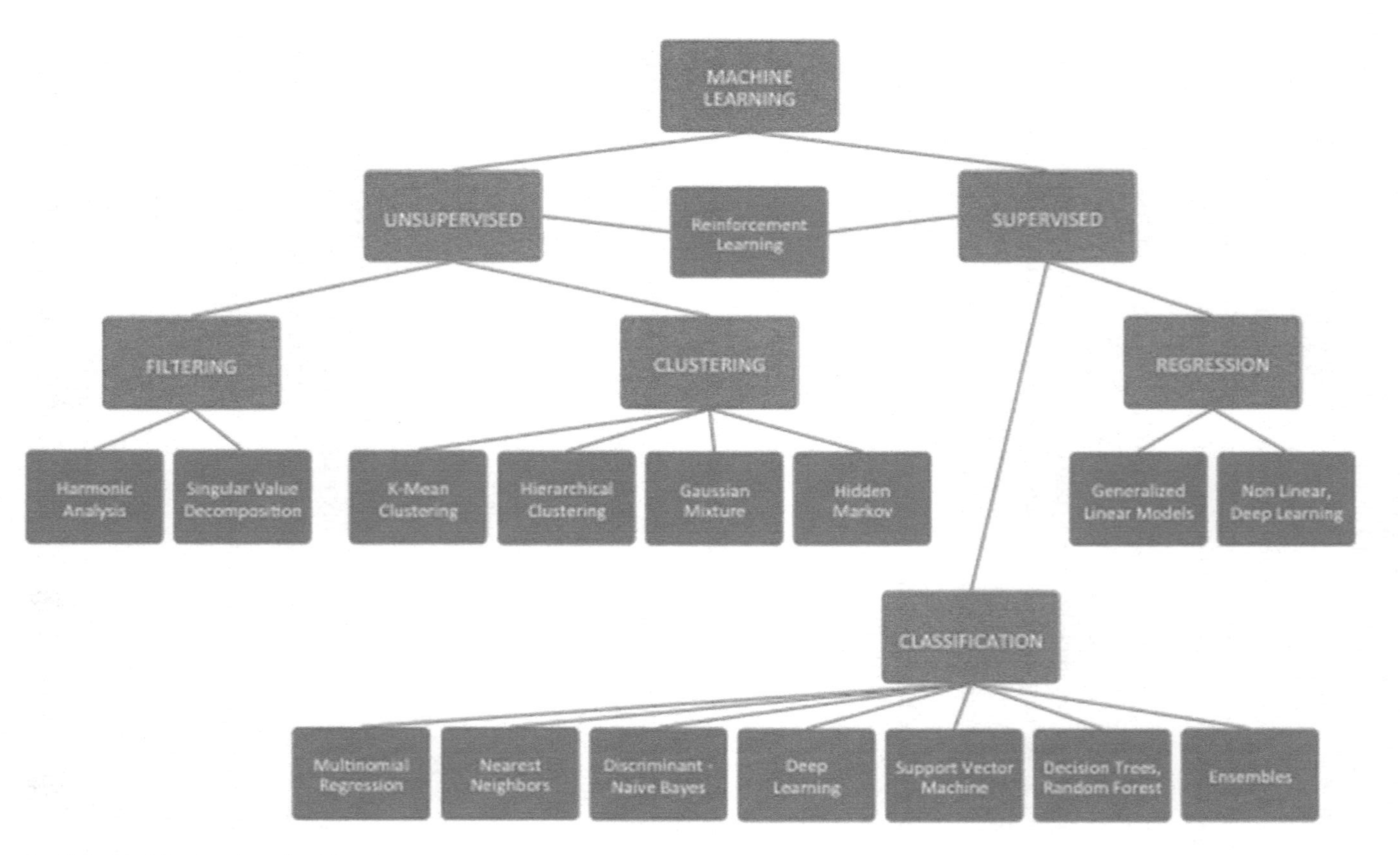

Fig. 7.1 Machine learning algorithms in common usage

Table 7.1 Comparison of machine learning algorithms in common usage [196]

Algorithm	Accuracy	Speed	Memory usage	Interpretability
Regression/GLM	Medium	Medium	Medium	High
Discriminant	Size dependent	Fast	Low	High
Naïve Bayes	Size dependent	Size dependent	Size dependent	High
Nearest neighbor	Size dependent	Medium	High	Medium
Neural network	High	Medium	High	Medium
SVM	High	Fast	Low	Low
Random Forest	Medium	Fast	Low	Size dependent
Ensembles	[a]	[a]	[a]	[a]

[a]The properties of Ensemble methods are the result of the particular combination of methods chosen by the user

the input variables (*features*) contain some categorical variables is also an important consideration. Not all methods can handle categorical variables, and some methods handle them better than others [174], see Table 7.1. More complex types of data, which are referred to as *unstructured* data (e.g. text, images, sounds), can also be processed by machine learning algorithms but they require additional steps of preparation. For example, a client might want to develop a predictive model that will learn customer moods and interests from a series of random texts sourced from both the company internal communication channel (e.g. all emails from all customers in past 12 months) and from publicly available articles and blogs (e.g. all newspapers published in the US in past 12 months). In this case, before using machine learning, the data scientist will use Natural Language Processing (NLP) to derive linguistic concepts from these corpora of unstructured texts. NLP algorithms are described in a separate section since they are not an alternative but rather an augmented version of machine learning, needed when one wishes to include unstructured data.

Considering the level of prior knowledge on the probability distribution of the features is essential to choosing the algorithm. All supervised machine learning algorithms belong to either one of two groups: parametric or non-parametric [165]:

- **Parametric learning** relies on prior knowledge on the probability distribution of the features. Regression Analysis, Discriminant Analysis and Naïve Bayes are parametric algorithms [165]. *Discriminant Analysis* assumes an independent Gaussian distribution for every feature (which thus must be numerical). *Regression Analysis* may implement different types of probability distribution for the features (which may be numerical and/or categorical). Some regression algorithms have been developed for all common distributions, the so-called *exponential family* of distributions. The exponential family includes normal, exponential, bi-/multi-nomial, χ-squared, Bernoulli, Poisson, and a few others. For the sake of terminology, these regression algorithms are called the *Generalized Linear Models* [165]. Finally, *Naïve Bayes* assumes independence of the features as in Discriminant Analysis but offers to start with any kind of prior distribution for the features (not only Gaussians) and computes their posterior distribution under the influence of what is learned in the training data.

- **Non-parametric learning does** not require any knowledge on the probability distribution of the features. This comes at a cost, generally in term of interpretability of the results. Non-parametric algorithms may generally not be used to explain the influence of different features relative to one another on the behavior of the response label, but still may be very useful for decision-making purposes. They include *K-Nearest Neighbor,* which is one of the simplest machine learning algorithms and where the mapping between features and response is evaluated based on a *majority-vote* like clustering approach. In short, each value of a feature is assigned to the label that is the most frequent (or a simple average if the label is numerical) across the cluster of k neighbor points, k being fixed in advance. Non-parametric algorithms in common usage also include *Neural Network*, *Support Vector Machine*, *Decision Trees/Random Forest*, and the customized *Ensemble* method which may be any combination of learning algorithms thereof [165, 195]. The advantages and limitations of these algorithms are summarized in Table 7.1.

Regression models have the broadest flexibility concerning the nature of variables handled and an ideal interpretability. For this reason, they are the most widely used [165]. They quantify the strength of the relationship between the response label and each feature, and together with the *stepwise regression* approach (which will be detailed below and applied in Sects. 7.5 and 7.6), they ultimately indicate which subsets of features contain redundant information, which features experience partial correlations and by how much [195].

In fact, regression may even be used to solve classification problems. *Logistic* regression and *multinomial* logistic regression [197] make the terminology confusing at first because these are types of regression that are classification methods indeed. The form of their inference function h maps a set of features into a set of *discrete* outcomes. In logistic regression the response is binary, and in multinomial regression the response may take any number of class labels [197].

So, why not *always* use a regression approach? The challenges start to surface when dealing with many features because in a regression algorithm some heuristic optimization methods (e.g. *Stochastic Gradient Descent*, *Newton Method*) are used to evaluate the relationship between features and find a solution (i.e. an optimal *weight* for every feature) by minimizing the loss function as explained in Sect. 6.1. Thus working with large datasets may decrease the robustness of the results. This happens because when the dataset is so large that it becomes impossible to assess all possible combinations of weights and features, the algorithm "starts somewhere" by evaluating one particular feature against the others, and the result of this evaluation impacts the decisions made in the subsequent evaluations. Step by step, the path induced by earlier decisions made, for example about the inclusion or removal of some given feature, may lead to drastic changes in the final predictions, a.k.a. *Butterfly effects*. In fact all machine learning algorithms may be considered simple conceptual departures from the regression approach aimed at addressing this challenge of robustness/reproducibility of results.

Discriminant analysis, *k*-nearest neighbor, and Naïve Bayes are very accurate for small datasets with a small number of variables but much less so for large datasets with many variables [196]. Note that discriminant analysis will be accurate only if the features are normally distributed.

Support vector machine (SVM) is currently considered the overall best performer, together with *Ensemble* learning methods [165] such as *bagged decision tree* (a.k.a. *random forest* [198]) which is based on a *bootstrapped* sampling[11] of trees. Unfortunately, SVM may only efficiently apply to classification problems where the response label takes exactly two values. Random forest may apply both to classification and regression problems, as for decision trees, with the major drawback being the interpretability of the resulting decision trees, which is always very low when there are many features (because the size of the tree renders big picture decisions impossible).

Until a few years ago, neural networks used to drain behind other more accurate and efficient algorithms such as SVM [165] or more interpretable algorithms such as regressions. But with recent increase of computational resources (e.g. GPU) [199] combined with recent *theoretical* development in Convolutional [200] (CNN, for image/video) and Recurrent [201] (RNN, for dynamic systems) Neural Network, Reinforcement Learning [202] and Natural Language Processing [203] (NLP, described in Sect. 7.4.3), Neural Networks have clearly come back [204, 205]. As of 2017 they are meeting unprecedented success and might be the most talked about algorithms currently in data science [203–205]. RNN are typically used in combination with NLP to learn *sequences* of words and recognize, complete and emulate human conversation [204]. The architecture of a general deep learning neural network and recurrent neural network are shown in Fig. 7.2. A neural network, to first approximation, can be considered a network of regression, i.e. multiple regressions of the same features complementing each other, and supplanted by regressions of regressions to enable more abstract levels of representations of the feature space. Each neuron is basically a regression of its input toward a 'hidden' state, its output, with the addition of a non-linear activation function. Needless to say, there are obvious parallels between the biologic neuron and the brain on one side, and the artificial neuron and the neural network on the other.

Takeaways – If you may remember only one thing from this section, it should be that regression methods are not always the most accurate but almost always the most interpretable because each feature will be assigned a weight with an associated p-value and confidence interval. If the nature of the variables and time permit, always give it a try to a regression approach. Then, use a table of pros and cons such as Table 7.1 to select *a few* algorithms. Because this is the beauty of machine learning: it is always worth trying more than one framework to evaluate the robustness of predictions, compare performances, and eventually build a compound model made of the best performers. At the end of the day, the Ensemble approach is by design as best as one may get with the available data.

[11] Bootstrap refers to successive sampling of a same dataset by leaving out some part of the dataset until convergence of the estimated quantity, in this case a decision tree.

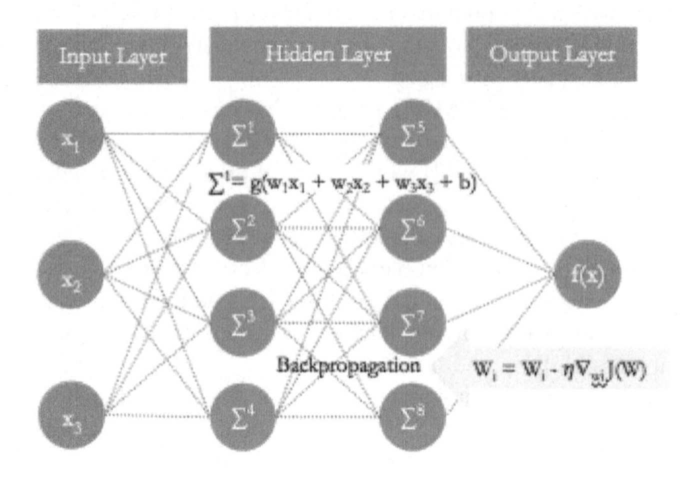

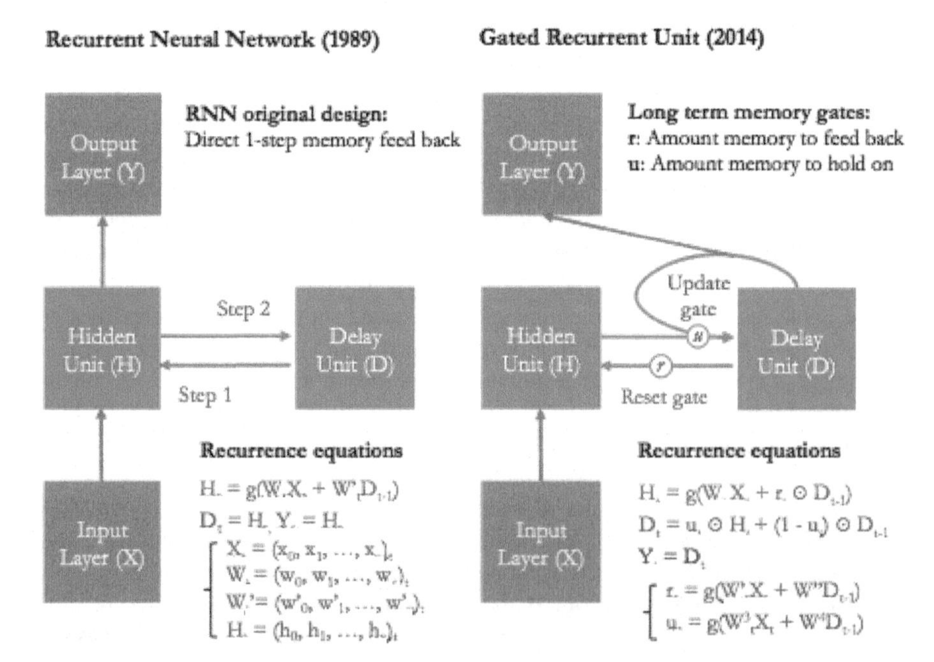

Fig. 7.2 Architecture of general (*top*) and recurrent (*bottom*) neural networks; GRUs help solve vanishing memory issues that are frequent in deep learning [205]

7.4.2 Model Design and Validation

Building and Evaluating the Model

Once an algorithm has been chosen and its parameters optimized, the next step in building up a predictive model is to address the *complexity tradeoff* introduced in Sect. 6.1 between under-fitting and over-fitting. The best regime between signal and noise is searched by *cross-validation*, also introduced in Sect. 6.1: a training set is defined to develop the model and a testing set is defined to assess its performance. Three options are available [165]:

1. **Hold-out**: A part of the dataset (typically between 60% and 80%) is randomly chosen to represent the training set and the remaining subset is used for testing
2. **K-folds**: The dataset is divided into *k* subsets. *k-1* of them are used for training and the remaining one for testing. The process is repeated *k* times, so that each fold gets to be the testing fold. The final performance is the average over the *k* folds
3. **Leave-1-out**: Ultimately *k*-folding may be reduced to leave-one-out by taking *k* to be the number of data points. This takes full advantage of all information available in the entire dataset but may be computationally too expensive

Model performance in machine learning corresponds to how well the hypothesis *h* in Eq. 7.7 may predict the response variable(s) for a given set of features. This is called the *error measure*. For classification models, this measure is the rate of success and failure (e.g. confusion matrix, ROC curve [206]). For regression models, this measure is the loss function introduced in Sect. 6.1 between predicted and observed responses, e.g. the Euclidean distance (Eq. 6.6). To change the performance of a model, three options are available [165]:

Option 1: **Add or remove some features by variance threshold of recursive feature selection**
Option 2: **Change the hypothesis function by introducing regularization, non-linear terms, or cross-terms between features**
Option 3: **Transform some features** e.g. by PCA or clustering

These options are discussed below, except for the *addition of non-linear terms* because this option requires deep human expertise and is not recommended given there exist algorithms that can handle non-linear functions automatically (e.g. deep learning, SVM). Deep learning is recommended for non-linear modeling.

Feature Selection

Predictive models provide an understanding of which variables influence the response variable(s) by measuring the strength of the relationship between features and response(s). With this knowledge, it becomes possible to add/remove features one at a time and see whether predictions performed by the model get more accurate and/or more efficient. Adding features one at a time is called *forward wrapping*, removing features one at a time is called *backward wrapping*, and both are called *ablative analysis* [165]. For example, stepwise linear regression is used to evaluate the impact of adding a feature (or removing a feature in backward wrapping mode) based on the p-value threshold 0.05 for a χ-squared test of the following hypothesis: *Does it affect the value of the error measure?*, where $H_1 = yes$ and $H_0 = no$. All these tests are done automatically at every step of the stepwise regression algorithm. The algorithm may also add/remove cross-terms in the exact same way. Ultimately, stepwise regression indicates which subsets of features contained redundant information and which features experience partial correlations. It selects features appropriately …and automatically!

Wrappers are perfect in theory, but in practice they are challenged by Butterfly effects when searching for the optimal weights of the features. That is, it is impossible to exhaustively assess all combinations of features. When the heuristic "starts somewhere" it impacts subsequent decisions made during the stepwise search, and certain features that might be selected in one search might be rejected in another where the algorithm starts somewhere else, and vice versa.

For very large datasets thus, a second class of feature selection algorithm may be used, referred to as *filtering*. Filters are less accurate than wrappers but more computationally effective and thus might lead to a better result when working with large datasets that prevent wrappers from evaluating all possible combinations. Filters are based on computing the matrix of correlations (Eq. 6.2) or associations (Eq. 6.4 or Eq. 6.5) between features, which is indeed faster than a wrapping step where the entire model (Eq. 7.7) is used to make an actual prediction and evaluate the change in the error measure. A larger number of combinations can thus be tested. The main drawback with filters is that the presence of partial correlations may mislead results. Thus a direct wrapping is preferable to filtering [165].

As recommended in Sect. 6.3, a smart tactic may be to use a filter at the onset of the project to detect and eliminate variables that are exceedingly redundant (too high ρ) or noisy (too low ρ), and then move on a more rigorous wrapper. Note another straightforward tactic here: when working with a regression model, the strength of the relationship between features relative to one another can be directly assessed by comparing the magnitude of their respective weights. This offers a solution for the consultant to expedite the feature selection process.

Finally, feature *transformation* and *regularization* are two other options that may be leveraged to improve model performance. Feature transformation builds upon the singular value decomposition (e.g. PCA) and harmonic analysis (e.g. FFT) frameworks described in Sect. 7.1. Their goal is to project the space of features into a new space where variables may be ordered by decreasing level of importance (please go back to Sect. 7.1 for details), and from there a set of variables with high influence on the model's predictions may be selected.

Regularization consists in restraining the magnitude of the model parameters (e.g. forcing weights to not exceed a threshold, forcing some features to drop out, etc) by introducing additional terms in the loss function used when training the model, or in forcing prior knowledge on the probability distribution of some features by introducing Bayes rules in the model.

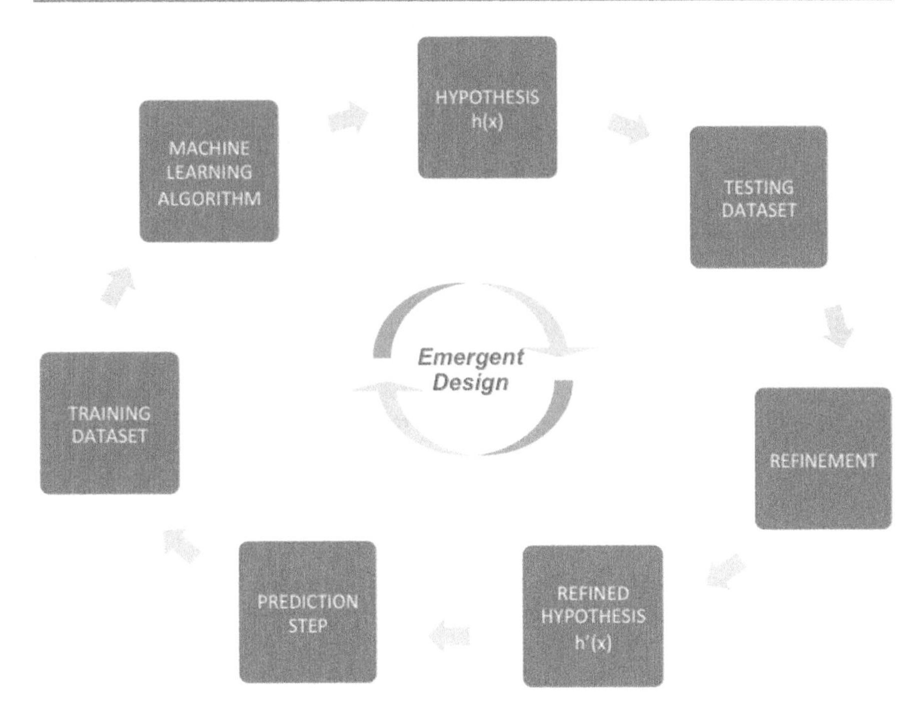

Fig. 7.3 Workflow of agile, emergent model design when developing supervised machine learning models

The big picture: agile and emergent design

The sections above, including the ones on signal processing and computer simulations, described a number of options for developing and refining a predictive model in the context of machine learning. If the data scientist, or consultant of the twenty-first century, was to wait for a model to be theoretically optimally designed before applying it, he could spend his entire lifetime working on this achievement! Some academics do. But this is not just an anecdote, as anyone may well spend several weeks reading through an analytics software package documentation before even starting to test his or her model. So here is something to remember: unexpected surprises may always happen, for any one and any model, when that model is finally used on real world applications.

For this reason, data scientists recommend an alternative approach to extensive model design: *emergent design* [207]. Emergent design does include data preparation phases such as exploration, cleaning and filtering, but quite precociously switches to building a working model and applying it to real world data. It cares less about understanding factors that might play a role during model design and more about the insights gathered from assessing real-world performance and pitfalls. Real-world feedbacks bring a unique value to orient efforts toward, for example, choosing the algorithm at the first place. Try one that looks reasonable, and see what the outputs look like — not to make predictions, but to make decisions about refining and improving performance (Fig. 7.3).

In other words, emergent design recommends to apply a viable model as soon as possible rather than to spend time defining the range of theoretically possible options. Build a model quickly, apply it to learn from real-world data, get back to model design, re-apply to real-world data, learn again, re-design and so forth. This process should generate feedbacks quickly with as little risks and costs as possible for the client, and in turn enable the consultant to come up with a satisfactory model in the shortest amount of time. The 80/20 rule always prevails.

7.4.3 Natural Language Artificial Intelligence

Let's get back to our example of a client who wishes to augment machine learning forecasts by leveraging sources of unstructured data such as customer interest expressed in a series of emails, blogs and newspapers collected over the past 12 months.

A key point to understand about *Natural Language Processing* (NLP) is that these tools often don't just learn by detecting signal in the data, but by associating patterns (e.g. subject-verb-object) and contents (e.g. words) found in new data with rules and meanings previously developed on prior data. Over the years literally, sets of *linguistic* rules and meanings known to relate to specific domains (e.g. popular English, medicine, politics) were consolidated from large collections of texts within each given domain and categorized into publicly available dictionaries called lexical corpora.

For example, the Corpus of Contemporary American English (COCA) contains more than 160,000 texts coming from various sources that range from movie transcripts to academic peer-reviewed journals, totaling 450 M words, pulled uniformly between 1990 and 2015 [208]. The corpus is divided into five sub-corpora tailored to different uses: spoken, fiction, popular, newspaper and academic articles. All words are annotated according to their syntactic function (part-of-speech e.g. noun, verb, adjective), stem/lemma (root word from which a given word derives, e.g. 'good', 'better', 'best' derive from 'good'), phrase, synonym/homonym, and other types of customized indexing such as time periods and collocates (e.g. words often found together in sections of text).

Annotated corpora make the disambiguation of meanings in new texts tractable: a word can have multiple meanings in different contexts, but when context is defined in advance, then the proper linguistic meaning can be more closely be inferred. For example the word "apple" can be disambiguated depending on whether it collocates more with "fruit" or "computer" and whether it is found in a gastronomy vs. computer related article.

Some NLP algorithms just parse text by removing *stop words* (e.g. white space) and standard suffixes/prefixes, but for more complex inferences (e.g. associating words with meaningful lemma, disambiguating synonyms and homonyms, etc), tailored annotated corpora are needed. The information found in most corpora relate to *semantics* and *syntax* and in particular sentence *parsing* (define valid grammatical constructs), *tagging* (define valid part-of-speech for each word) and *lemmatization* (rules that can identify synonyms and relatively complex language morphologies). All these together may aim at inferring name entity (is apple the fruit, computer or firm), what action is taken on these entities, from whom/what, with which intensity, intent, etc. Combined with sequence learning (e.g. recurrent neural networks introduced in Sect. 7.4.1), it enables to follow and emulate speech. And combined with unsupervised learning (e.g. SVD/PCA to combine words/lemma based on their correlation), it enables to derive high-level concepts (named *latent semantic analysis* [209]). All these are at the limit of the latest technology of course, but we are getting to a time when most mental constructs that make sense to a human can be encoded indeed, and thereby when artificially intelligent designs may emulate human behavior.

Let us take a look at a simple, concrete example and its algorithm in details. Let us assume we gathered a series of articles and blogs written by a set of existing and potential customers on a company, and want to develop a model that identifies the sentiment of the articles for that company. To make it simple let us consider only two outcomes, positive and negative sentiments, and derive a classifier. The same logic would apply to quantify sentiment numerically using regression, for example on a scale of 0 (negative) to 10 (positive).

1. Create an annotated corpus from the available articles by tagging each article as 1 (positive sentiment) or 0 (negative sentiment)
2. Ensure balanced classes (50/50% distribution of positive and negative articles) by sampling the over-represented class
3. For each of the n selected articles:
   - Split article into a list of words
   - Remove stop words (e.g. spaces, and, or, get, let, the, yet,…) and short words (e.g. any word with less than three characters)
   - Replace each word by its base word (lemma, all lower case)
   - Append article's list of lemma to a master/nested list
   - Append each individual lemma to an indexed list (e.g. dictionary in Python) of *distinct* lemma, i.e. append a lemma only when it has never been appended before
4. Create an $n \times (m + 1)$ matrix where the $m + 1$ columns correspond to the m distinct lemma + the sentiment tag. Starting from a null-vector in each of the n rows, represent the n articles by looping over each row and incrementing by 1 the column corresponding to a given lemma every time this lemma is observed in a given article (i.e. observed in each list of the nested list created above). Each row now represents an article in form of a frequency vector
5. Normalize weights in each row to sum to 1 to ensure that each article impacts prediction through its word frequency, not its size
6. Add sentiment label (0 or 1) of each article in last column
7. Shuffle the rows randomly and hold out 30% for testing
8. Train and test a classifier (any of the ones described in this chapter, e.g. logistic regression) where the input features are all but the last column, and the response label is the last column

9. The classifier can now be used on any new article to indicate whether the article's sentiment for the company is positive or negative, with level of confidence identified in step 8
10. If a logistic regression was used, the coefficients of the features (i.e. lemma) can be screened to identify words that have most positive and most negative sentiments in each article

Note in the simple NLP exercise above we annotated ourselves the training data by flagging which articles/blogs had positive vs. negative sentiment, yet we still relied on existing corpora in step 3 to remove stop words and, more importantly, identify lemma. Lemmatizers can be found easily (e.g. NLTK [210]) hence no project in NLP ever really starts from scratch, for the benefit of everyone working with artificial intelligence.

Naturally, we could have leveraged existing sentiment analysis APIs [211] to flag the training dataset in the first place, instead of doing it manually. But NLP is an emerging science, and existing annotated corpora are often not specific enough for most projects. NLP projects often require to manually develop and extend corpora by defining new or alternative rules specifically for the project. Often both the entities to analyze (e.g. customer name, article ID) and the measure on them (e.g. sentiment for a given company) need be coded into linguistic concepts before analysis. Note *Latent Semantic Analysis* (i.e. SVD on sets of lemma [209]) can be used to code a complex concept as a weighted average of multiple lemma.

By transforming a corpus of text into linguistic concepts, NLP enables unstructured data to be transformed into what is effectively structured data, i.e. a finite set of features that take on numeric value for each customer. As we saw, these features can then be processed by usual machine learning algorithms.

Similar to the conceptual transition we make from unsupervised to supervised learning when the goal moves from detecting patterns to predicting outcomes, with NLP we make another transition: one from a purely data-driven machine learning approach to an approach where both data and pre-defined knowledge drive the learning process. This concept of *reinforcement learning* [202] is the state of the art, and recently met highly publicized success: when Google DeepMind beat the Go world champion in March 2016 [212], it was through a gigantic reinforcement learning process that took place prior to the tournament. Reinforcement learning consists in defining rules based on prior or new knowledge from the environment and adding rewards on partial outcomes, *dynamically* as the learning proceeds, when these partial outcomes can be identified as closer to the final goal (e.g. winning the game). A set of rules is referred to as a policy, a partial outcome as a state and the model as an agent for sequential decision making. The learning process consists in interacting with the environment and exploring as many states as possible to improve the agent's policy. The policies are applied through state-specific terms in the overall loss function, which can be propagated to the weights of the agent's features by standard minimization (e.g. Gradient Descent). This is like adding religion to science, adding supervision on the go based on ad-hoc rules coming from the environment, observed state, actions and reward, to a more formal learning process. Google developed reinforcing policies specifically for Go by having the computer play against itself over multiple months before the tournament. When the computer faced the world champion Lee Sedol on March 2016, it won four games to one. Five months earlier, it had beaten Fan Hui, the European champion, five games to zero [213]. A year later, it's upgraded version was reported to beat its earlier version by 100 to nothing. Go allows for more potential moves to play than there are atoms in the universe, and is claimed to be the most complex game known to mankind. Now of course, playing games is not all what makes someone, or something, intelligent.

7.5 Case 1: Data Science Project in Pharmaceutical R&D

The following example presents a data science project that was carried out for a pharmaceutical company within a 2-week time frame. It applies machine learning in the context of feature selection and predictive modeling to identify *biomarkers* that have potential for reducing R&D operation costs. This example was chosen because it goes beyond the type of ubiquitous market research introduced at the beginning of this chapter for which abundant examples can already be found online, and fully illustrates the breadth of potential applications beyond its own R&D operation contexts (e.g. market research, credit rating).

The Challenge

In this project, machine learning is applied to analyze the Parkinson's Progression Markers Initiative dataset (PPMI [214]) developed by the *Michael J. Fox* foundation, which provides longitudinal data on the largest sample to-date of patients with Parkinson disease (as of 2016). The cohort contains hundreds of patient with and without Parkinson (PD vs. HC patients) who were followed over several years. The goal of this project is to identify clinical features that associate with measurements of Dopamine transporters (DScan) decrease in the brain: Dscans are accurate but expensive (>$3000/scan) and thus finding cheaper alternative biomarkers of Parkinson progression would lead to dramatically reduced cost for our client and its healthcare partners, who regularly diagnose and follow patients with or at risk of Parkinson.

The Questions

What are the clinical features that correlate with Dscan decrease?
Can we raise a predictive model of Dscan decrease?
If yes, what are its essential features?

The Executive Summary

Using a 4-steps protocol: 1- Exploration, 2- Correlations, 3- Stepwise Regression and 4- Cross-Validation with a HC-PD Classification Learning Algorithm, three clinical features were found to be important for predicting Dscan decrease: the *Hoehn and Yahr* motor score, the *Unified Parkinson Disease Rating* score when measured by physicians (not when self-administered by patients) and the *University of Pennsylvania Smell Identification* score. Our client may use these three features to predict Parkinsonians with >95% success rate.

Exploration of the Dataset

The total number of patients available in this study between year 0 (baseline) and year 1 was 479, with the Healthy Cohort (HC) represented by 179 patients and the cohort with Parkinson Disease (PD) represented by 300 patients.

Discard features for which > 50% of patients' records are missing – A 50% threshold was applied to both HC and PD cohorts for all features. As a result, every feature containing less than 90 data points in either cohort was eliminated.

Discard non-informative features – Features such as enrollment date, presence/absence to the various questionnaires and all features containing only one category were eliminated.

As a result of this cleaning phase, 93 features were selected for further processing, with 76 features treated as *numericals* and 17 features (Boolean or string) treated as *categoricals.*

Correlation of Numerical Features

The correlation ρ between each feature and the Dscan decrease was computed, and consistency was evaluated by computing the correlation ρ' with the HC/PD label. In Table 7.2, features are ranked by descending order of magnitude of the correlation coefficient. Only the features for which the coefficient has a p-value < 0.05 *or* a magnitude >0.1 are shown.

The features within the dotted square in Table 7.2 are the ones for which the correlation with Dscan decrease $\rho > 0.2$ with a p-value < 0.05. These were selected for further processing.

The features for which the cross-correlation (Fig. 7.4) with some other feature was >0.9 with a p-value < 0.05 were considered redundant and thus non-informative, as suggested in Sect. 6.3. For each of these cross-correlated groups of features, only the two features that had the highest correlation with DScan decrease were selected for further processing; the others were eliminated.

As a result of this pre-processing phase based on correlation coefficients, six numerical features were selected for further processing: the Hoehn and Yahr Motor Score (NHY), the physician-led Unified Parkinson Disease Rating Score (NUPDRS3), the physician-led UPenn Smell Identification Score (UPSIT4) and the Tremor Score (TD).

Selection of Numerical Features by Linear Regression

Below are the final estimates after a stepwise regression analysis (introduced in Sect. 7.4) using the p-value threshold 0.05 for the χ-squared test of the change in the sum of squared errors $\sum_{i=1}^{479}(y_i - \hat{y}_i)^2$ as criterion for adding/removing features, where y and $\hat{y}$ are the observed and predicted values of DScan respectively for each patient.

Feature	Θ	95% Conf. interval		p-value
NHY	−0.112	−0.163	−0.061	1.81e−05
NUPDRS3	−0.010	−0.016	−0.005	0.001
UPSIT4	0.011	0.002	0.019	0.021
NHY:NUPDRS3	0.007	0.004	0.010	4.61e−06

Table 7.2 Correlation of features with DScan (*left*) and HC/PD label (*right*)

Correlation with Dscan			Correlation with HC / PD		
Feature	**ρ**	**p-value**	**Feature**	**ρ'**	**p-value**
NHY	-0.32	1.73e-12	NHY	0.88	1.25e-154
UPSIT4	0.30	1.51e-11	NUPDRS3	0.81	1.90e-112
UPSIT total	0.29	6.87e-11	NUPDRS total	0.79	2.20e-102
NUPDRS3	-0.29	1.24e-10	TD	0.69	1.60e-67
NUPDRS total	-0.27	3.02e-09	UPSIT total	-0.66	1.16e-60
UPSIT1	0.26	4.05e-09	UPSIT1	-0.62	2.71e-52
UPSIT2	0.24	7.12e-08	NUPDRS2	0.61	8.74e-51
UPSIT3	0.24	7.87e-08	UPSIT4	-0.60	1.82e-47
TD	-0.22	8.41e-07	UPSIT3	-0.58	1.40e-44
NUPDRS2	-0.21	5.05e-06	UPSIT2	-0.57	5.92e-43
PIGD	-0.16	0.00061	PIGD	0.47	1.42e-27
SDM total	0.15	0.00136	NUPDRS1	0.32	4.49e-13
SFT	0.15	0.00151	SCOPA	0.31	7.35e-12
RBD	-0.13	0.00403	SDM1	-0.29	1.16e-10
pTau 181P	0.13	0.00623	SDM2	-0.29	1.16e-10
SDM1	0.12	0.00674	SDM total	-0.28	4.12e-10
SDM2	0.12	0.00674	RBD	0.26	1.06e-08
WGT	-0.11	0.01537	STAI1	0.23	1.78e-07

NHY Hoehn and Yahr Motor Score, *NUPDRS-x* Unified Parkinson Disease Rating Score (the numbers *x* correspond to different conditions in which the test was taken, e.g. physician-led vs. self-administered), *UPSIT-x* University of Pennsylvania *Smell* Identification Test, and *TD* Tremor Score

Two conclusions came out of this stepwise regression analysis: First, TD is not a good predictor of DScan despite the relatively high correlation with the HC/PD label found earlier (Table 7.2). It was verified that a strong outsider data point exists that explains this phenomena. Indeed, when this outsider (shown in Fig. 7.5) is eliminated from the dataset, the original correlation ρ' of TD with the HC/PD label drops significantly.

Secondly, the algorithm suggests that a cross-term between NHY and NUPDRS3 will improve model performance. At this stage thus, three numerical features and one cross-term were selected: NHY, NUPDRS3, UPSITBK4 and a cross-term between NHY and NUPDRS3.

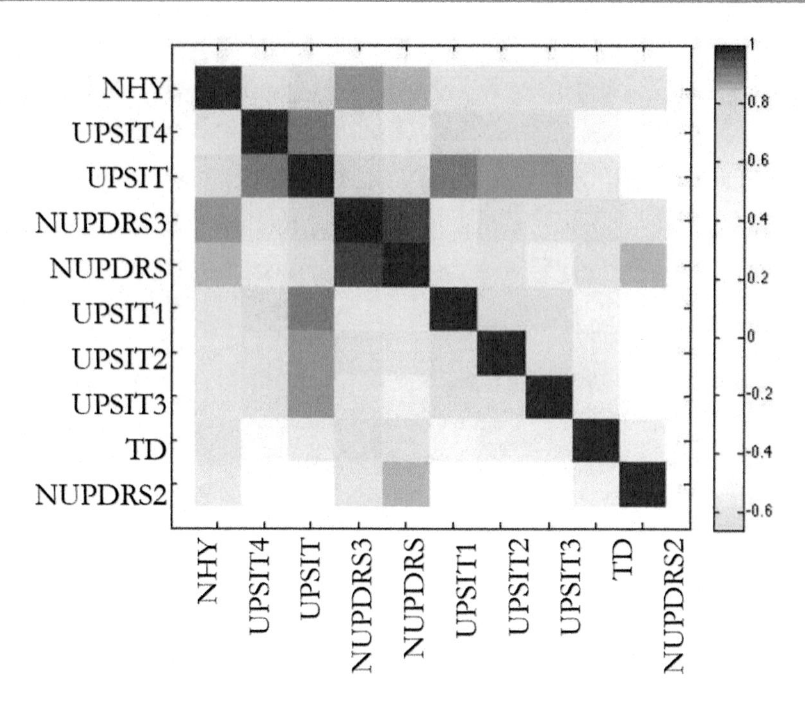

Fig. 7.4 Cross-correlation between features selected in Table 7.2

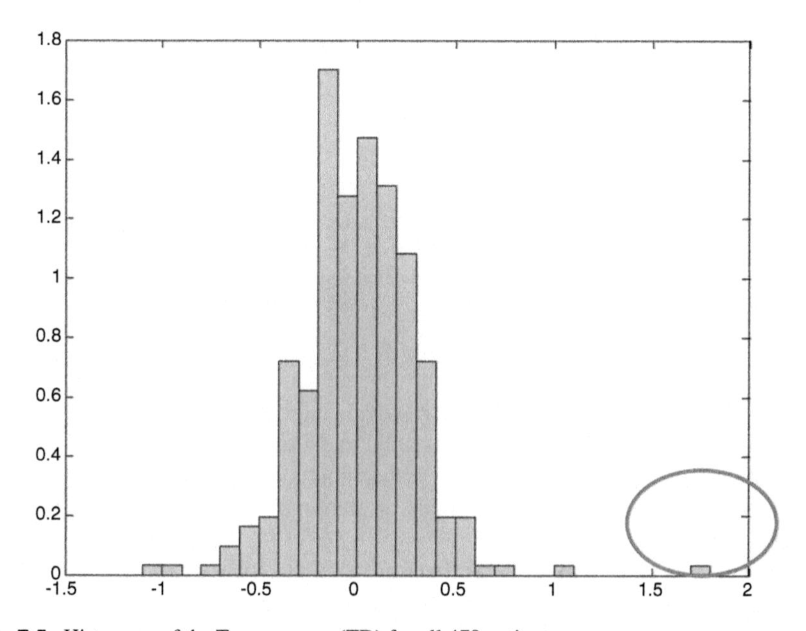

Fig. 7.5 Histogram of the Tremor score (TD) for all 479 patients

Table 7.3 Evaluation of sample size bias for categorical variables in the HC and PD labels

Feature	Label	HC (179)	PD (300)
Race	BLACK	8 (5%)	3 (1%)
Psychiatric	TRUE	0	5 (2%)
RB Disorder	TRUE	36 (20%)	119 (40%)
Neurological	TRUE	13 (7%)	149 (50%)
Skin	TRUE	25 (14%)	35 (12%)

Selection of Categorical Features by Linear Regression

Below are the final estimates after stepwise regression using the same criterion as above for adding/removing features, but performed on a model's starting hypothesis containing only the categorical features.

Feature	Θ	95% Conf. interval		p-value
RB disorder	−0.033	−0.075	0.012	0.137
Neurological	−0.072	−0.118	−0.031	0.001
Skin	0.091	0.029	0.1523	0.004

Two features referred to as *Psychiatric (positive)* and *Race (black)* were also suggested by the algorithm to contribute significantly in the model's hypothesis function, but looking at the cumulative distribution of these two features in the HC and PD labels (see below), it was concluded that both signals result from a fallacy of small sample size: the feature *Psychiatric* contains only five instances, all in PD. The feature *Race* contains only 5% of HC instances and 1% of PD instances. Both were considered non-significant (too small sample) and thereby eliminated.

As a result of this stepwise regression analysis for categorical variables, three categorical features were selected: the *REM Sleep Behavior Disorder* (RBD), the *Neurological* disorder test, and the *Skin* test. The value of the feature *Skin* was ambiguous at this stage: it did not seem to significantly associate with PD according to Table 7.3 (14% vs. 12%), yet the regression algorithm suggested that it could improve model performance. The feature *Skin* was given a benefit of doubt and thereby conserved for further processing.

Predictive Model of Dopamine Transporter Brain Scans

Below are the final estimates after stepwise regression using the same criterion as above for adding/removing features, but performed on a model's starting hypothesis containing both the numerical and categorical features selected in the previous steps above.

Feature	Θ	p-value	Θ	p-value
NHY	-0.061	0.012	-0.060	0.010
NUPDRS3	-0.0001	0.908	-0.0002	0.900
UPSIT4	0.014	0.002	0.015	0.001
RB Disorder	-0.002	0.925	*eliminated*	-
Neurological	-0.0003	0.990	*eliminated*	-
Skin	0.066	0.030	0.066	0.026

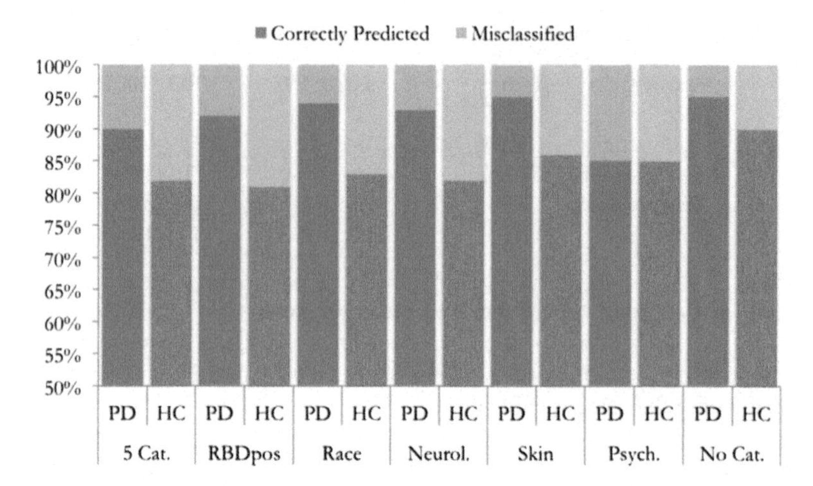

Fig. 7.6 Confusion matrix for a HC-PD machine learning classifier based on logistic regression with different hypothesis functions $h(x)$

The final model's hypothesis suggested by the algorithm does not contain any cross-term nor NUPDRS3 which has lost significance relative to NHY and UPSIT4 (both in term of weight and p-value, see above).

The same applied for the two categorical features, *RB sleep disorder* and *neurological disorder*, which have relatively small weights and high p-values.

Finally, the feature *Skin disorder* remained with a significant p-value and is thus a relevant predictor of DScan. It was not communicated to the client as a robust predictor however, because there is no association with the HC and PD labels as noted earlier (Table 7.3).

In conclusion, the *Hoehn and Yahr Motor Score* (NHY) and the physician-led *UPenn Smell Identification Score* (UPSIT4) are the best, most robust predictors of DScan decrease in relation to the Parkinson disease. A linear regression model with the two features NHY and UPSIT4 is thereby a possible predictive model of DScan decrease in relation to the Parkinson disease.

Cross-validation 1: Comparison of Logistic Learning Models with Different Features

The predictive modeling analysis above identified a reduced set of three clinical features that may be used to predict DScan (NHY, UPSIT4 and eventually NUPDRS3). None of the five categorical features (Psychiatric, Race, RB Disorder, Neurological and Skin) was selected as a relevant predictor of DScan with statistical significance.

A HC-PD binary classifier was developed to cross-validate these conclusions made on the basis of DScan measurements by predicting the presence/absence of the Parkinson disease as effectively diagnosed. This HC-PD classifier was a machine learning logistic regression with 60% training hold out that included either the five categorical features, one of these five features, or none of these five features.

From Fig. 7.6, which shows the rates of successes and failures for each of the seven machine learning classification algorithms tested, we observe that using all

Table 7.4 Comparison of the error measure over tenfolds for different machine learning classification algorithms

Algorithm	20 features	3 features
Discriminant analysis	0.019	0.006
k-nearest neighbor	0.382	0.013
Support vector machine	0.043	0.010
Bagged tree (random forest)	0.002	0.002

five categorical features as predictors of HC vs. PD gives the worst performance, and using no categorical predictor (using only the three numerical features NHY, UPSIT4 and NUPDRS3) is similar to or better than using any one of these categorical predictors. Thereby, we confirmed that none of the categorical features may improve model performance when trying to predict whether a patient has Parkinson.

Cross-validation 2: Performance of Different Learning Classification Models
To confirm that using the three clinical features NHY, UPSIT4 and NUPDRS3 is sufficient to raise a model of DScan measurements, the performance of several machine learning classification modeling approaches that aim at predicting the presence/absence of the Parkinson disease itself was compared with each other. In total, four new machine learning models were built, each with a 60% training hold out followed by a ten-fold cross validation. These four models were further compared when using all 20 features that ranked first in term of marginal correlation in Table 7.2 instead of only the three recommended features, see Table 7.4.

From Table 7.4 which shows the average mean squared error over ten folds of predictions obtained with each of the four new machine learning classification algorithms, we observe that using the three features NHY, UPSIT4 and NUPDRS3 appears sufficient and optimum when trying to predict whether a patient has Parkinson.

From Fig. 7.7, which shows the rate of successes and failures for each of the five machine learning classification algorithms tested (includes logistic regression), we confirm again that NHY, UPSIT4 and NUPDRS3 are sufficient and optimum when trying to predict whether a patient has Parkinson.

General Conclusion – Three clinical features were identified that may predict DScan measurements and thereby reduce R&D costs at the client's organization: the Hoehn and Yahr motor score, the Unified Parkinson Disease rating score, and the UPenn Smell Identification test score. These three features perform similar or better compared to when using more of the features available in this study. This conclusion was validated across a variety of learning algorithms developed to predict whether a patient has Parkinson. SVM and Random Forest perform best but the difference in performance was non-significant (< 2%), which supports the use of a simple logistic linear regression model. The latter was thus recommended to the client because it is the easiest for all stakeholders to interpret.

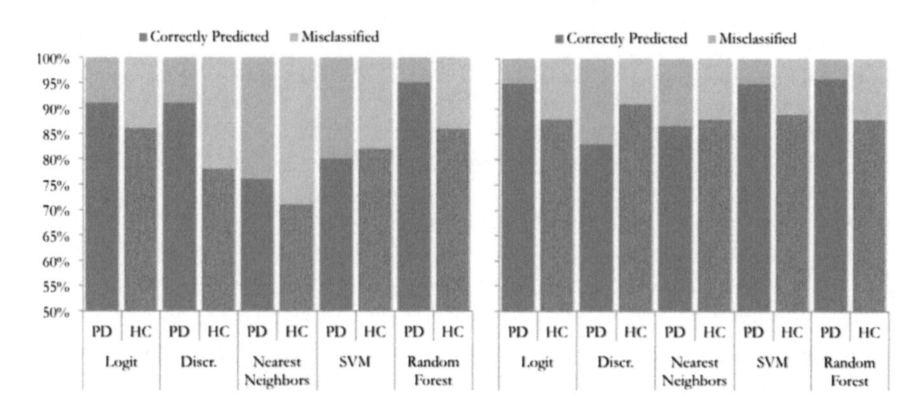

Fig. 7.7 Comparison of the confusion matrix for different machine learning classification algorithms using 20 features (*left*) and 3 features (*right*)

7.6 Case 2: Data Science Project on Customer Churn

This second example presents a data science project that was also carried out within a 2-week time frame, for a consulting engagement at one of the top tier management consulting firms. It applies machine learning to identify *customers who will churn*, and aims at extracting both quantitative and qualitative recommendations from the data for the client to make proper strategic and tactical decisions to reduce churn in the future. This example was chosen because it starts delving into many of the typical subtleties of data science, such as lack of clear marginal correlation for any of the features chosen, highly imbalanced dataset (90% of customers in the dataset do not churn), probabilistic prediction, adjustment of prediction to minimize false positives at the expense of false negatives, etc.

The Challenge

In this project, the client is a company providing gas and electricity who has recently seen an increase in customer defection, a.k.a. churn, to competitors. The dataset contains hundreds of customers with different attributes measured over the past couple of months, some of whom have churned, some have not. The client also provided a list of specific customers for which we are to predict whether each is forecasted to churn or not, and with which probability.

The Questions

Can we raise a predictive model of customer churn?
What are the most explicative variables for churn?
What are potential strategic or tactical levers to decrease churn?

The Executive Summary

Using a 4-step protocol: 1- Exploration, 2- Model design, 3- Performance analysis and 4- Sensitivity analysis/interpretations, we designed a model that enables our client to identify 30% of customers who will churn while limiting fall out (false positive) to 10%. This study supports short term tactics based on discount and longer term contracts, and a long term strategy based on building synergy between services and sales channels.

Exploration of the Dataset

The exploration of the dataset included feature engineering (deriving 'dynamic' attributes such as weekly/monthly rates of different metrics), scatter plots, covariance matrices, marginal correlation and Hamming/Jaccard distances with are loss functions designed specifically for binary outcomes (see Table 7.5). Some key issues to be solved were the presence of many empty entries, outliers, collinear and low-variance features. The empty entries were replaced by the median for each feature (except for features with more than 40% missing in which case the entire feature was deleted). The customers with outlier values beyond six standard deviations from the mean were also deleted.

Some features, such as prices and some forecasted metrics, were collinear with $\rho > 0.95$, see Fig. 7.8. Only one of each was kept for designing machine learning models.

Table 7.5 Top correlation and binary dissimilarity between top features and churn

Top pearson correlations with churn		
Feature	Original	Filtered
Margins	0.06	0.1
Forecasted meter rent	0.03	0.04
Prices	0.03	0.04
Forecasted discount	0.01	0.01
Subscription to power	0.01	0.03
Forecasted consumption	0.01	0.01
Number of products	−0.02	−0.02
Antiquity of customer	−0.07	−0.07

Top binary dissimilarities with churn		
Feature	Hamming	Jaccard
Sales channel 1	0.15	0.97
Sales channel 2	0.21	0.96
Sales channel 3	0.45	0.89

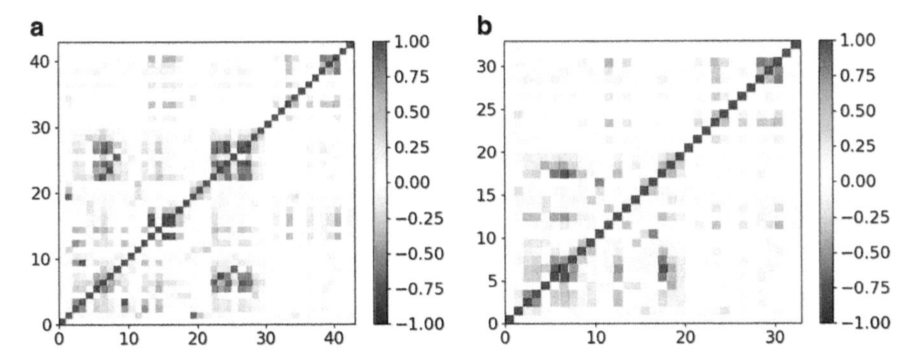

Fig. 7.8 Cross-correlation between all 48 features in this project (*left*) and after filtering collinear features out (*right*)

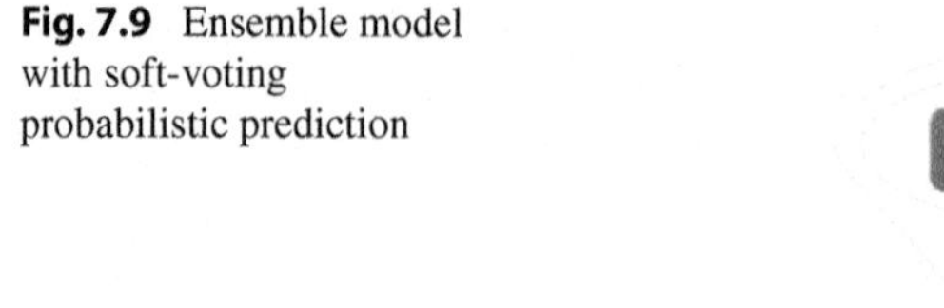

Fig. 7.9 Ensemble model with soft-voting probabilistic prediction

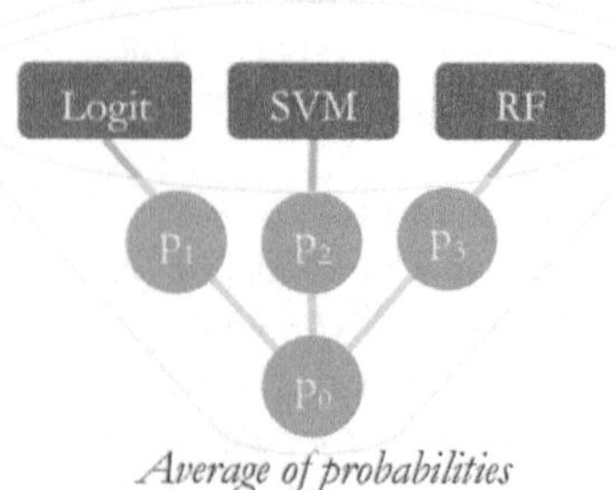

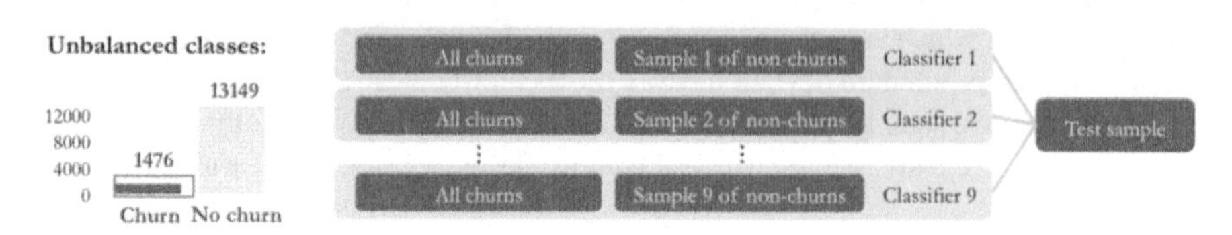

Fig. 7.10 Distribution of churn and no-churn customers and nine-fold ensemble model designed to use all data with balance samples and improve generalization

Model Design

A first feature selection was carried out before modeling using a variance filter (i.e. removed features with no variance in more than 95% of customers) and later on during modeling by stepwise regression. Performance of learners was assessed by both 30% hold-out and ten-fold cross validation. A logistic regression, support vector machine and a random forest algorithm were trained both separately and in ensemble as depicted in Fig. 7.9. The later is referred to as 'soft-voting' in the literature because it takes the average probability of the different models to classify customer churn.

Clients with churn represented only 10% of the training data. Two approaches were tested to deal with class imbalance. Note for both, performance was evaluated on a test set where class imbalance was maintained to account for real world circumstances.

- **Approach 1** (baseline): Training based on random sampling of the abundant class (i.e. clients who didn't churn) to match the size of the rare class (i.e. client who did churn)
- **Approach 2:** Training based on an ensemble of nine models using nine random samples of the abundant class (without replacement) and a full sample of the rare class for each. A nine-fold was chosen because class imbalance was 1–9, it is designed to improve generalization, see Fig. 7.10.

Performance Analysis

The so-called receiver-operator characteristic (ROC [206]) curve is often used to complement contingency tables in machine learning because it provides an

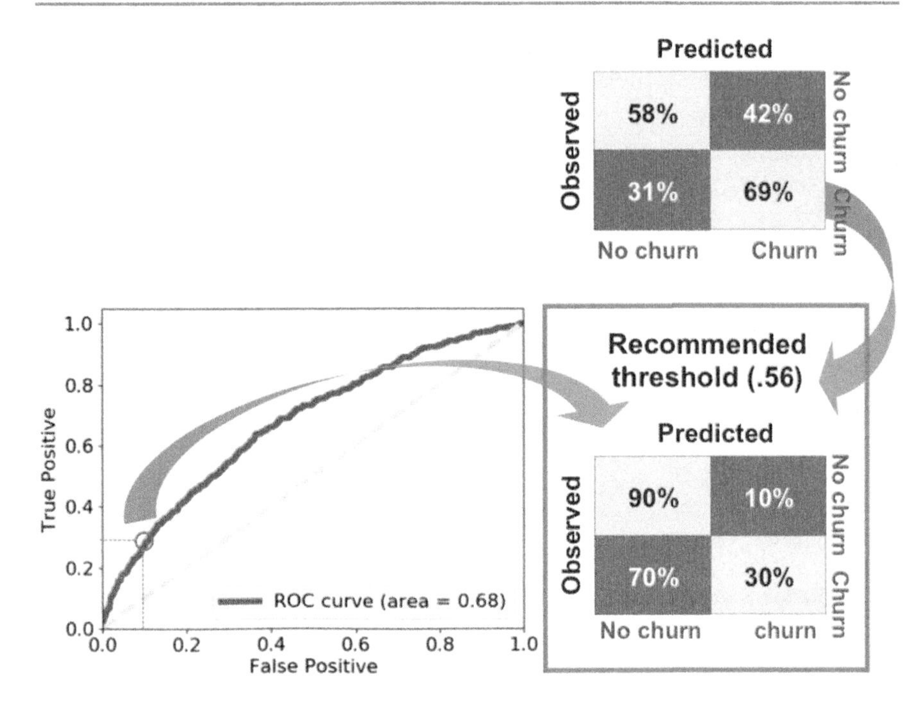

Fig. 7.11 Performance of the overall best learner (nine-fold ensemble of random forest learners) and optimization of precision/fall out

invariant measure of accuracy under changing the probability threshold for inferring positive and negative classes (in this project churn vs. no churn respectively). It consists in plotting all predictions classified positive accurately (a.k.a. true positives) vs. the ones misclassified as positives (a.k.a. false positive or fall out). The probabilities predicted by the different machine learning models are by default calibrated so that $p > 0.5$ correspond to one class, and $p < 0.5$ corresponds to the other class. But of course this threshold can be fine tuned to minimize misclassified instances of one class at the expense of increasing misclassification in the other class. The area under the ROC curve is an elegant measure of overall performance that remains the same for any threshold in use. From Table 7.6, we observe that random forest has the best performance, and that the nine-fold ensemble does generalize a little better, with a ROC AUC score of 0.68. This is our 'best performer' model. It enables to predict 69% of customers who will churn, although with significant fall out of 42% when using a probability threshold of 0.5, see Fig. 7.11. Looking at the ROC curve, we see that the same model can still predict 30% of customers, a significant proportion for our client, albeit this time with a highly minimized fall out of just 10%. Through grid search, I found this threshold to be $p = 0.56$. This is the model that was recommended to the stakeholders (Table 7.6).

Table 7.6 Performance of the learning classifiers, with random sampling of the rare class or nine-fold ensemble of learners, based on Accuracy, Brier and ROC measures

Performance measures			
Algorithm	Accuracy	Brier	ROC
Logit	56%	0.24	0.64
Stepwise	56%	0.24	0.64
SVM	57%	0.24	0.63
RF	**65%**	**0.24**	**0.67**
Ensemble	61%	0.23	0.66
9-logit ensemble	55%	0.26	0.64
9-SVM ensemble	61%	0.25	0.63
9-RF ensemble	**70%**	**0.24**	**0.68**
9-ensemble ensemble	61%	0.25	0.65

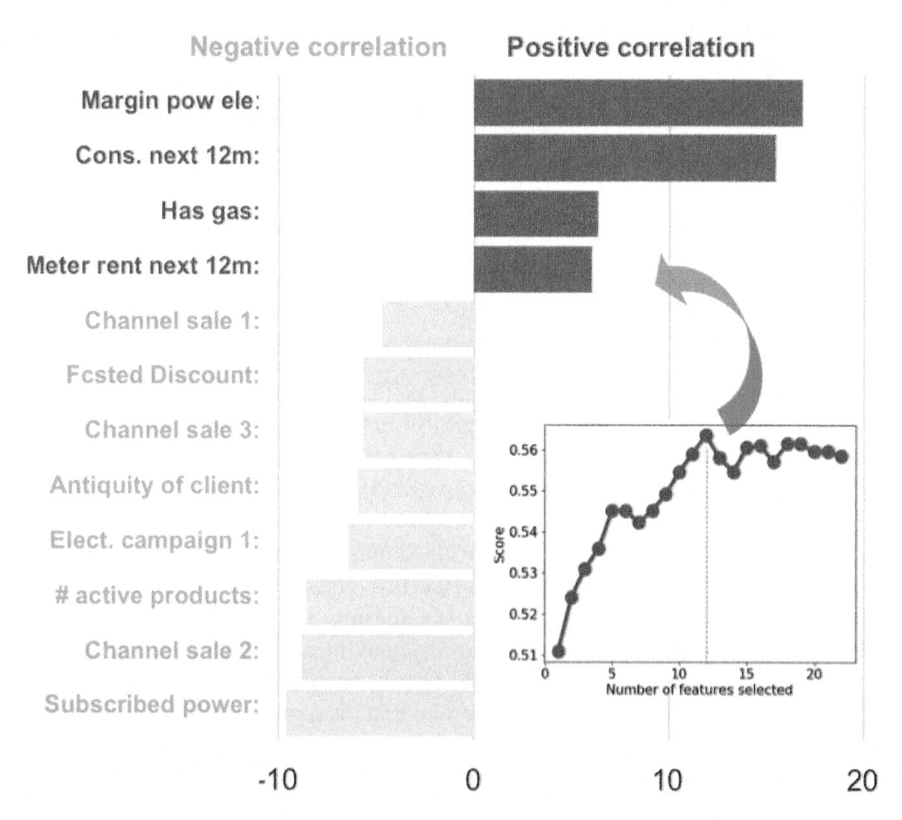

Fig. 7.12 Explicative variables of churn identified by stepwise logistic regression (*close-up*) and ranked by their relative contribution to predict churn

Sensitivity Analysis and Interpretations

A recurrent feature selection carried out by stepwise logistic regression lead to the identification of 12 key predictors, see Fig. 7.12. Variables that tend to induce churn were high margins, forecasted consumptions and meter rent. Variables that tend to

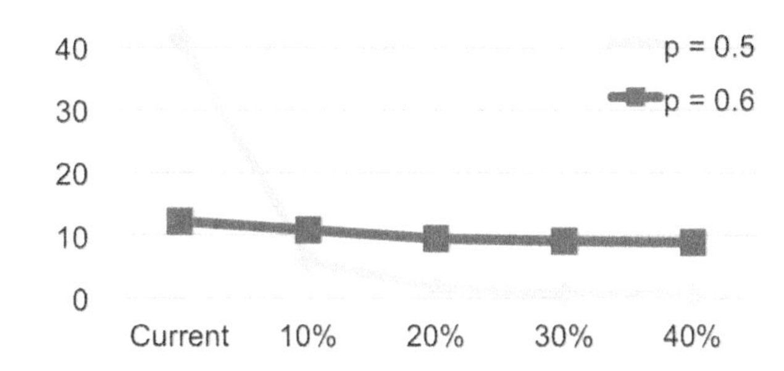

Fig. 7.13 Sensitivity analysis of discount strategy: proportion of customer churn predicted by logit model after adding a discount for clients predicted to churn

reduce churn were forecasted discount, number of active products, subscription to power, antiquity (i.e. loyalty) of the customer and also three of the sales channels. Providing a discount to clients with high propensity to churn seems thus a good tactical measure, and other strategic levers could also be pulled: synergies, channels and long term contracts.

Finally, a sensitivity analysis was carried out by applying an incremental discount to clients identified by learners, and then re-running the models to evaluate how many clients were still predicted to churn with that discount level. If we consider the model with minimized fall out (because client had expressed interest in minimizing fall out), our analysis predict that a 20% discount will reduce churn significantly (i.e. by 25%) with minimal fall out (10%). Given the true positive rate with this threshold is about 30%, we can safely forecast that the discount approach will eliminate at least 8% of churn. See Fig. 7.13 for details. This is the first tactical step that was recommended to the stakeholders.

Principles of Strategy: Primer

8

A competitive strategy augments a corporate organization with inherent capabilities to sustain superior performance on a long-term basis [74]. Many strategy concepts exist and will be described in this chapter, with a special focus placed on practical matters such as key challenges and "programs" that can be used as roadmaps for implementation. Popular strategies include *the five forces, the value chain, the product life cycle, disruptive innovation* and *blue ocean*, to name a few. The reader is invited to consider these models as a simple aid to thinking about reality, since none of these theoretical concepts or authors thereof ever claim to describe the reality for any one particular circumstance. They claim to facilitate discussion and creativity over a wide range of concrete business issues. Armed with such tools, the consultant may examine whether his/her client does indeed enjoy a competitive advantage, and develop a winning strategy.

8.1 Definition of Strategy

Defining a strategy enables executive managers to rationalize their resource allocation and decision-making. To be of assistance, the consultant needs to determine a set of objectives together with a comprehensive set of actions, resources and processes that are required to meet these objectives.

Let us thus start this chapter on strategy by defining a high level categorization [215] based on the nature of the consultant's objectives:

- **Functional level:** Which activities should the client engage in?
- **Business level**: What scope is needed to satisfy market demand?
- **Corporate level**: What businesses should the client invest in?

These three categories will be used in the next chapter to organize the different concepts of advanced strategy. Of course the merit of defining such categories is again purely pedagogic: developing a strategy should always involve holistic

J. D. Curuksu, *Data Driven*, Management for Professionals,
https://doi.org/10.1007/978-3-319-70229-2_8

approaches that appreciate the many potential interactions and intricacies within and beyond the proposed set of activities. At the end of the day indeed, these subtleties are what make the difference between tactic and strategy.

Harvard Professor Michael Porter defined strategy in three points [74]:

- Create a unique and valuable position involving a different set of activities
- Choose what not to do – competing in a marketplace involves making trade-offs because not all activities may be compatible. For example, a business model that works well for offering high end products to a niche market generally doesn't work well for offering low cost products to a mainstream market
- Create fit between the proposed activities and the client's internal capabilities. Most activities interact with each other and thus must reinforce each other

8.2 Executing a Strategy

A strategy can only be as good as the information available. Business decisions are frequently made in a state of uncertainty, and thus the outcomes of a strategy are never quite as expected. At first, a strategy may appear as a logic and analytical process by which to understand the environment, the industry, the client's strengths, weaknesses, competitors, and required competencies. This kind of diligence is essential to deploy a competitive edge, but as time passes the various uncertainties will start to be resolved and existing strategies will need to be modified accordingly [5].

A strategy can also only be as good as its execution. So, assuming an impeccable strategy, how to execute it nicely? This passage from theory to practice starts with articulating milestones and objectives that relate the strategy to the client's mission, vision and values. A set of programs should be described, together with organizational structures, guidelines, policies, trainings, control and information sharing systems. Warning signals should be monitored and the client should be ready to adapt the original strategy as events unfold. The objectives may remain the same, but the strategies by which these objectives are to be achieved should be subject to continuous review. In such a context, the relationship between a consultant and a client will extend beyond the theory. Both partners shall engage in perpetual innovations.

8.3 Key Strategy Concepts in Management Consulting

8.3.1 Specialization and Focus

Michael Porter referred to as *generic strategy* [216] the approach adopted by a corporate organization to face the inevitable dilemma between pursuing premium-based differentiation or price-based specialization. He also recognized a key dependency on the level of *focus*, i.e. which segment of customers the organization focuses on. Since then many authors brought additional support to Porter's underlying claim that all corporate organizations in the long term have to "choose" between offering a price advantage or offering a quality premium to its customers, and that pursuing both directions at the same time is doomed to fail [100].

The *premium-based differentiation strategy* involves making a product appear different in the mind of the customers along dimension others than price. A product may be perceived more attractive because of better quality or technical features, but also because of better design, customer services, reliability, convenience, accessibility and combinations thereof [19]. Promotional efforts and brand perception also represent powerful sources of premium-based differentiation that have nothing to do with enhancing product quality or technical features. A more detailed discussion can be found in section 8.4. Finally, note that many cases have been reported where the simple fact of increasing the price could enhance customer's perception of quality, but this marketing strategy has more to do with psychology than business management.

The *price-based differentiation* and cost-leadership strategies are closely related to the concepts of *learning curve* [217] and *economy of scale*. These two concepts are theoretical efforts aimed at quantifying and predicting the benefits associated with cumulative experience and production volume. For example, the learning curve predicts that each time the cumulative volume of production in a manufacturing plant doubles (starting when the plant opens its doors), the cost of manufacturing per unit falls by a constant and predictable percentage [15]. Of course, generalizing the learning curve to every company and circumstance would be absurd, but as witnessed by an impressive amount of cases over the past 50 years, the learning curve theory appears surprisingly accurate in a great variety of circumstances. The "learning" efficiencies can come from many sources other than economy of scale such as labor efficiency (e.g. better skilled workers, automation), technological improvement, standardization, new product, new business model, new market focus [15].

Market focus is an omnipresent, generic and constantly evolving strategy that all organizations must pursue if they are to establish a competitive position and remain competitive in their marketplace [100]. Focusing on well-defined customer segments is necessary to understand well-defined needs and circumstances that have potential to foster interest in the organization's value proposition. The better a company understands and solves specific customer problems, the more competitive it becomes.

8.3.2 The Five Forces

To first approximation, the level of attractiveness and competition in an industry may be framed within the 5-force analysis scheme innovated in 1979 by Harvard Professor Michael Porter [218]. The underlying idea is that an organization should look beyond its direct competitors when analyzing its competitive arena. According to Porter, four additional forces may hurt the organization's bottom line (Fig. 8.1). Savvy customers can force down prices by playing the organization and its rivals

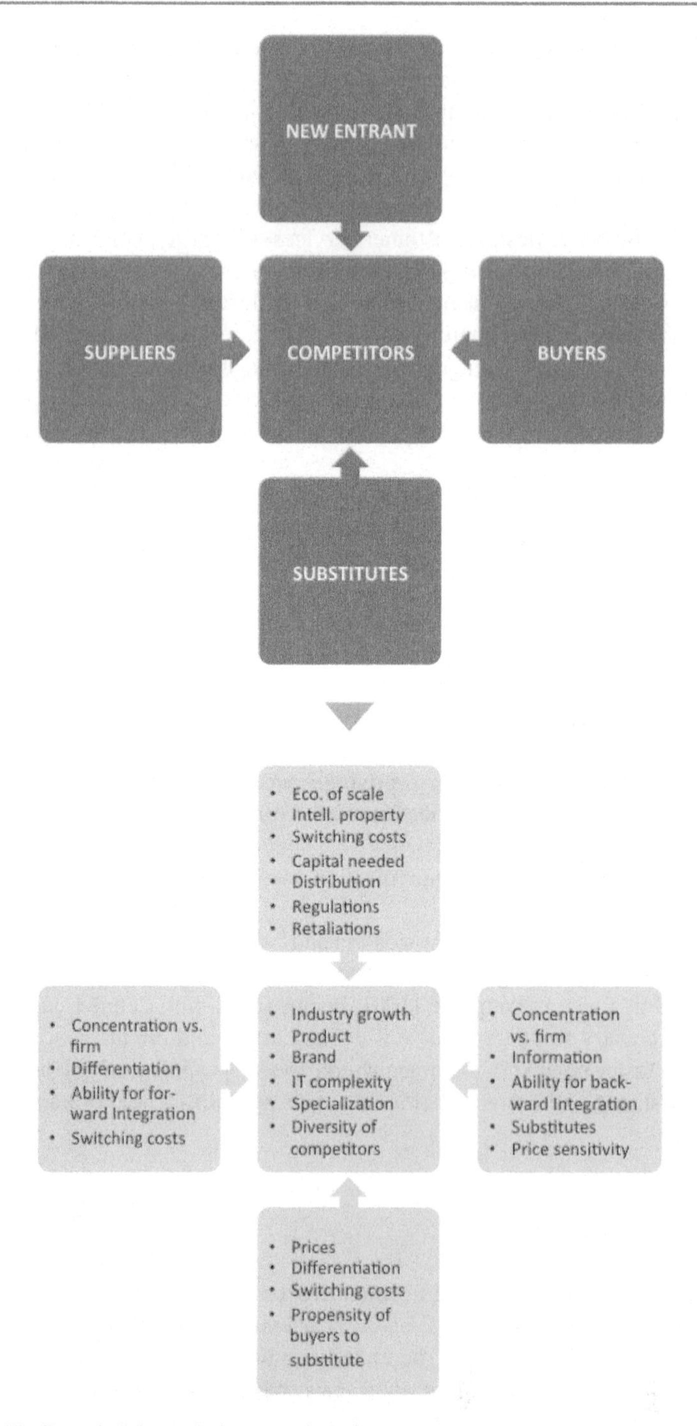

Fig. 8.1 The Porter's 5-forces industry analysis framework

against one another. Powerful suppliers can constrain the organization's profits by charging higher prices. Substitute offerings can lure customers away. And new entrants can leverage disruptive innovations and ratchet up the investment required for an incumbent to maintain its business model.

Prof. Porter later completed this theory by recognizing the need for factoring-in the influence of additional components such as complementary offering, innovation, government and inter-dependency between market players [17]. Still, the 5-force analysis scheme remains an efficient starting point when the consultant needs to analyze a market.

8.3.3 The Value Chain and Value Network

The chain of activities that a product passes through from creation to delivery in an organization (Fig. 8.2) is called the value chain [17]. It is so named because each activity is supposed to serve a purpose, and ultimately this purpose is always to add value to the product being offered by the organization. An organization should perpetually look for maximizing the added value and minimizing the cost of each activity in its value chain, as well their inter-relationships, in accordance with their specialization strategy (differentiation or cost leadership). A value chain can be optimized by benchmarking best practices (i.e. learn from the most successful players), but innovating new technologies and business models (see disruptive innovation and blue ocean below) can sometimes avoid interference from competitors altogether and thereby produce enormous returns.

Ultimately, a value chain is embedded in a value *network* [18, 75] that extends upstream into its suppliers and downstream into its consumers (the vertical dimension), radiates through similarly positioned players (the horizontal dimension) and naturally gives rise to growth strategies referred to as vertical and horizontal integrations [100].

The revolution in the economics of information brought about with the Internet has fundamentally changed the nature of corporate success when it pertains to value networks. The phenomenon has been referred to as *deconstruction* [219], where modular organizations regularly emerge and leverage each other at virtually every link of the value network, as made possible by easier access to rich and consistent information across the network of suppliers and consumers [220]. This information is not anymore the rare currency that it used to be in the pre-Internet era. Without this powerful lever that used to give highly integrated organizations an enormous competitive advantage, alternative levers—outsourcing, partnership, coexistence, sharing—have become the hallmark of modern businesses.

This hallmark is no universal law of course: depending on the clients and circumstances under consideration, vertical integration may still well provide a key strategic advantage.

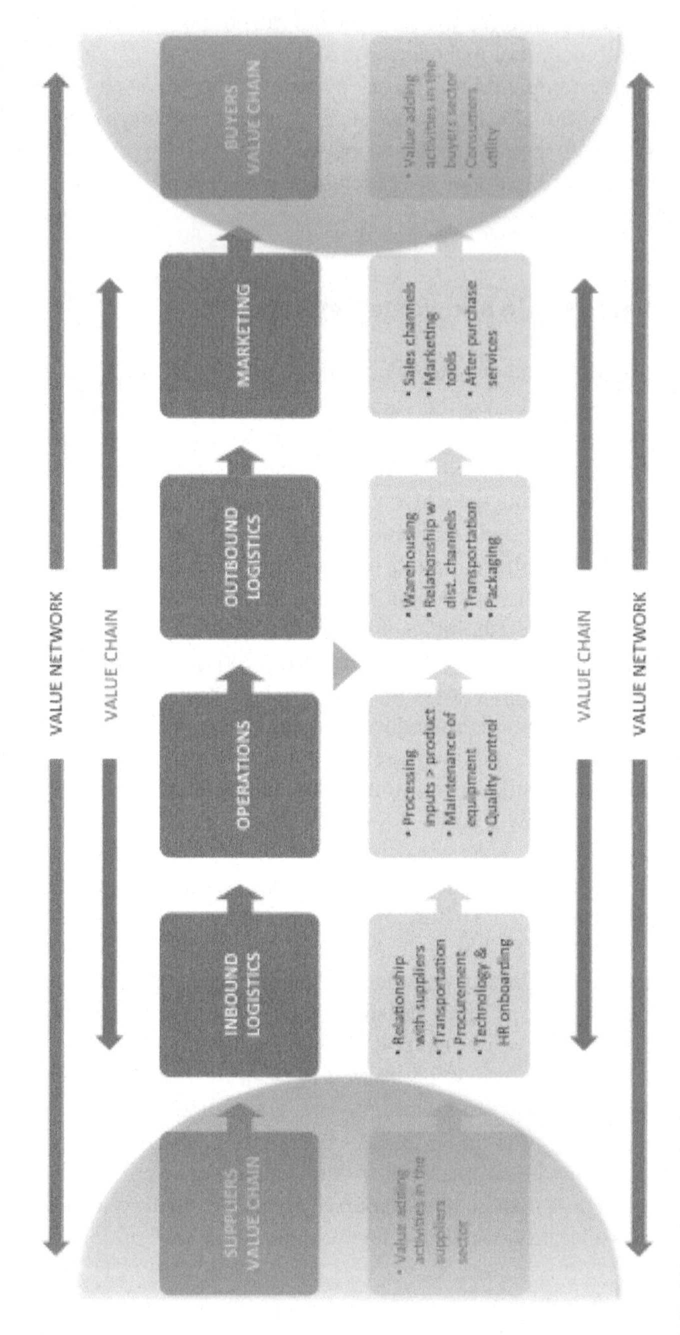

Fig. 8.2 Value chain and value network

8.3.4 Integration

Integration may occur upstream or downstream in the value network that a company finds itself in (vertical integration), or at a same level (horizontal integration).

Horizontal integration is a typical example of *Mergers and Acquisitions* (M&A, section 9.3.3). Such integration may serve very diverse purposes. Common purposes are: developing markets (increased penetration or new markets), developing products (line extension or diversification), responding to a competitive threat, creating synergies (accelerate the learning or best practice process) and procuring tax advantages.

Vertical integration extends the value chain upstream or downstream, which is referred to as backward and forward integration respectively. As mentioned above, deconstruction and modularization is on the rise in many marketplaces. This trend may be seen as the inverse of vertical integration, but it has obvious limits: the very purpose of any organization is to bring a set of resources/processes together, i.e. to *integrate* value-adding activities.

Moreover, deconstruction results in structural changes [219] that are not always associated with significant changes in consumption. Indeed, from the consumer perspective a fully integrated corporation or an organization loaded with partners at every link of its value chain does not necessarily produce any perceivable difference. For example, that Starbucks relies on partnership for bringing coffee beans into its facilities, or Pfizer on academia for bringing-in new lead projects in its laboratories, changes nothing in the consumer experience. The consumer still looks for the next nearest coffee place when in need of a delicious espresso and for the next nearest pharmacy when in need of a prescription refill.

In contrast to the consumer perspective, from the corporate organization perspective vertical integration is a subtler cost/benefit analysis problem that has been completely transformed by the revolution in the economics of information [220]. The past 20 years have brought the emergence of a multitude of *navigators*, in every marketplace, most often Internet-enabled service based agencies that help organizations cope with the complexity of doing business in a deconstructed world [221].

Integration may represent both a threat and an opportunity in the eyes of a corporate organization, as it gives birth to a new cocktail of bargaining leverage between suppliers, buyers and the organization respective to one another. Buying a powerful supplier or insourcing its activities (backward integration) may remove restraints and unleash new potentials. But it may as well bring unanticipated challenges and competitive pressure, affect core competencies and deteriorate brand image. More generally, integration often presents benefits that consultants should weigh against risks. On the side of benefits, forward and horizontal integrations enable the client to enter new markets and backward integration to secure inputs at lower cost. On the side of risks, horizontal integration calls for competitive response and diminishes focus; vertical integration increases risk of exposure associated with entering a new industry.

Vertical integration, most notably, can change the logic of entire value networks and ecosystems in which organizations live in. This is often ill perceived by the consultant's client [219]. The consultant in contrast, can more easily engage into industry-wide, big-picture, unbiased analyses.

8.3.5 Portfolio Strategies

In finance, the portfolio theory frames most investment decisions [222]. It is embolden in the simple idea that investing in several stocks is *always*[1] more profitable in the long run. A simple mathematical proof exists, and is explained here. A portfolio is visualized by positioning every stock that an organization/individual owns in a 2D diagram, where one axis represents expected return and the other axis represents expected risks. The total expected return (and total expected risk) is not only a function of each stock's return (and risk respectively), it also depends on a correlation factor that characterizes the evolution of *every pair of stocks*. For n stocks, the number of such factors is $!n$ (which is defined as $!n = n \times (n-1) \times (n-2) \times (\ldots) \times 1$). Because $!n >> n$, investing in multiple stocks brings opportunities (and threats) that reach far beyond the landscape of *individual stocks*, because the growth outcome of a given portfolio where n is large is mainly driven by stock-correlations and not by each stock's individual growth. To understand this, consider a portfolio made of 10 stocks ($n = 10$). For $n = 10$,$!n = 3.6$ M and thus the aggregated growth of each individual stock represents less than 0.001% of all factors that drive the growth of the entire portfolio! 99.99% of the growth is driven by correlations *between* stocks.

The concept of portfolio theory in finance is useful to understand the concept of portfolio strategy in management consulting. Indeed, different companies have come up with different portfolio management approaches, yet the mathematical theory developed in the finance world provides a common underlying framework to all these management strategies (Fig. 8.3).

Portfolio strategies address the question of which products and businesses should a company continue or start to invest in, and which ones should the company sell [84]. Three popular portfolio management strategies are the BCG, McKinsey and ADL matrices [100]. These three matrices differ by which factors are believed to have the most influence on portfolio management and decision-making. They also differ by their level of granularity (BCG's has 4 entries, McKinsey's has 9 entries, ADL's has 16 entries). But again, all these techniques bear in common that they rely on two variables that are simple extensions of the expected return and the expected risk variables defined in the unified (mathematical) theory of finance introduced earlier …into the more *qualitative* world of business management.

The BCG matrix considers relative market share and market growth as key drivers of tactical decisions (with two levels in each dimension, resulting in a 4-quadrant matrix), the McKinsey matrix considers relative market position and industry attractiveness (for example as defined based on Porter's 5 forces introduced above), and the ADL matrix considers competitive position and industry maturity. Needless to say, these pairs of variables differ more in the details of how each company initially

[1] Except in circumstances of nearly *perfect* correlation between different stock.

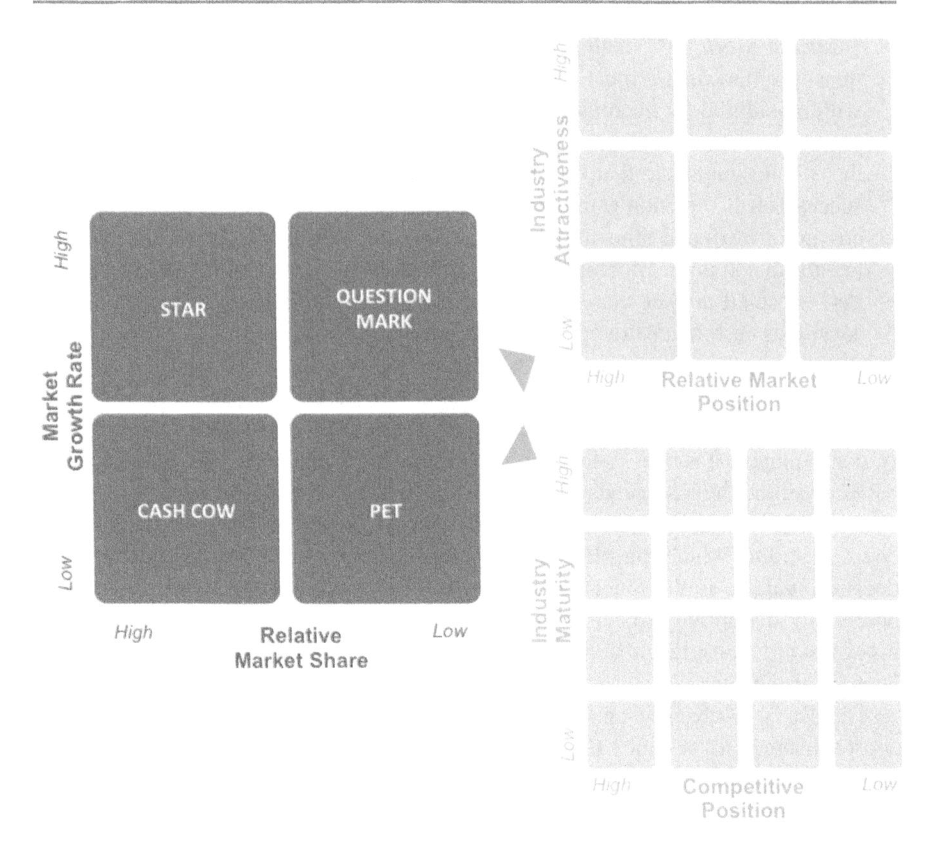

Fig. 8.3 The portfolio matrix (left: BCG, top: GE/McKinsey, bottom: AD Little)

defined them that in any fundamental way, and thereby of how each consultant will redefine them in practice. Market share, market position and competitive position all have to do with company's strengths/weaknesses. Market growth, industry attractiveness and industry maturity all have to do with opportunities/threats in a given sector.

The McKinsey and ADL matrices also differentiate in that they were originally defined on the basis of strategic business units (SBU [223]) rather than individual products, but the BCG matrix applies to SBU and individual product equally well.

Let us look in more details at one of these matrices, the BCG matrix [84], for which the four quadrants have become popular business classes:

- ***Star*** – A high market share business in a high growth industry, requiring savvy management to maintain the competitive edge
- ***Cash cow*** – A high market share business in a low growth industry. If there is no significant growth perspective in this profitable business, its returns are best used to fund the cash needs of other, more innovative projects (the *question marks*) in order to ensure long term growth

- ***Question mark*** – A small market share business in a high growth industry. A question mark is a project with low return relative to stars and cash cows, but with a potential for becoming a star. These are the innovative projects of tomorrow's business. They are sometimes referred to as *problem children* because there is no guarantee that investing in these projects will lead them to become successful. If they don't, they will become *pets*. As described below when discussing growth and innovation, several question mark projects should regularly be introduced in the pipeline to maintain healthy prospects of future growth
- ***Pet*** – A small market share business in a low growth industry. A pet essentially consumes cash in an attempt to remain competitive

With the concept of *Product Life Cycle* in mind (described in the section *Market Analysis* below), it is easy to understand that all Stars will, eventually, stop growing. In consequence, a success sequence across the BCG matrix is one where today's cash cows fund the cash needs of question marks, today's question marks eventually transition into stars, and today's stars (unfortunately) eventually transition into cash cows. Deciding which question mark projects to invest in is key to success. This decision-making is complicated by the risks and opportunities associated with potentially disruptive projects that might look like pets at first, until one day an innovator redefines the game of competition. *Disruptive innovations* are introduced later in this chapter.

The BCG, McKinsey and ADL matrices illustrate how the portfolio matrix approach can enjoy a wide range of applications: the consultant may adapt the two variables that account for client strength and market opportunity to the specifics of each case and each client. At the end of the day, a portfolio matrix becomes a customized tool aimed at conveniently visualizing the position of each product/business relative to one another. As the cousin theory in the finance world teaches us, the future of a portfolio, and thus the client organization, is far more driven by the inter-relationships and synergies (*covariances* in mathematical parlance) between the different businesses that the client may pursue than by the growth of any individual business.

8.3.6 Synergy

The portfolio theory illustrates that the fit between different activities in a company is very important, and that these inter-relationships may be quantified through finance theory. Defining and quantifying synergies is particularly important for high-investment operations such as mergers and acquisitions. Merger and acquisition targets may be valuated through more or less rigorous equations (e.g. Net Present Value (NPV) and Price Earning (p/e) ratio, respectively), and integrating synergies in these equations has a potential to significantly change the results. For example, one might add an effective term for *discounted synergies* [224] in the NPV equation, and treat synergies in the same way as any other tangible *asset*. Whichever valuation method is used (NPV or p/e ratio), synergies should always be carefully scrutinized.

Without moving into the details of quantitative valuation, it is useful to familiarize with the different dimensions in which an organization may develop or improve synergies. Synergies between two businesses may be divided in four categories [100]:

- **Market Synergies**, when one business leverages the customer base, distribution channels or brand identification already in place in another business
- **Product Synergies**, when one business extends the product line of another business, fills-in excess production capacity left over by the other business, increases the bargaining power with suppliers (i.e. improves procurement), or leverages employee functions already in place in the other business (e.g. accountants, engineers)
- **Technological Synergies**, when one business improves value or reduces cost of some activities in its value chain by bringing in operational technologies implemented in another business (e.g. process design, equipment, transportation, IT platform)
- **Intangible Synergies**, when managerial know-how and business model components are effectively transferred from one business to another

8.3.7 The *Ansoff* Growth Matrix

Corporate growth strategies are innumerable, diverse, and context-dependent. Theories may however be efficiently used as "roadmaps" to navigate between different "categories" of growth strategies. These roadmaps are essential to the consultant in order to walk its client through a relatively exhaustive palette of options before a decision is made to commit to just one or a few, potentially high stake and high risk, strategies.

Beside the notion of *organic* vs. *inorganic* growth, which essentially relates to whether expansion is undertaken internally or through M&A, the *Ansoff matrix* [153] is an elegant way to segment growth strategies into high-level categories.

Let us start from the beginning. The goal of all corporate endeavors, one shall agree, is embolden in the concept of value chain: for every activity undertaken, a company attempts to maximize value and minimize cost, which is turn maximizes value for stakeholders (e.g.: customers, consumers, shareholders, society). As such, a growth strategy is always an attempt at delivering better products at lower cost, which in turn increases purchases. But because purchase levels can be increased with a number of strategies that relate to stakeholder accessibilities and perceptions (e.g. promotions, distribution channels, pricing) and not actual characteristics of the products, Igor Ansoff defined two routes for corporate expansion [153]: product development and market development (Fig. 8.4).

Most of the options that an organization may consider for growth can be visualized into one of the quadrants of the Ansoff matrix. They either relate to penetration (selling more of the same products in the same markets), market expansion (selling

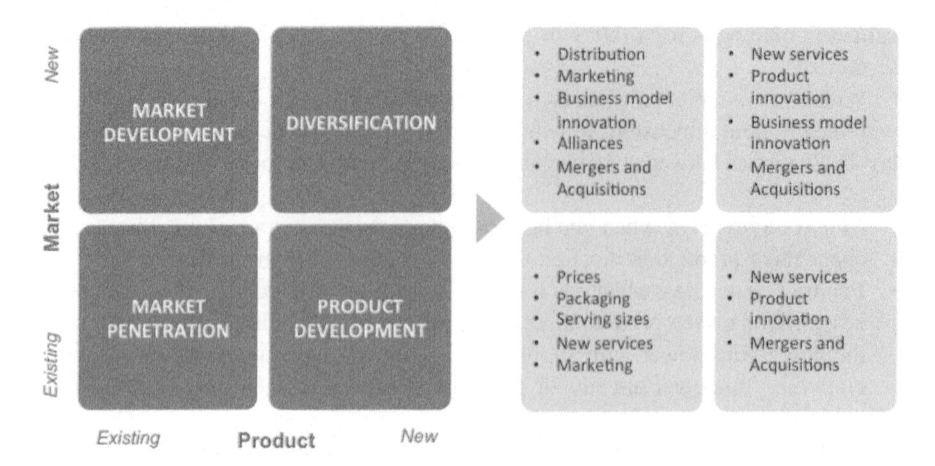

Fig. 8.4 Menu of strategic options in each quadrant of the Ansoff matrix

the same products in new markets), product diversification (selling new products in the same markets), or unrelated diversification (selling new products in new markets). As a rule of thumb, unrelated diversification is often the most complex and unpredictable route for expansion. Priority should thus be given to strategies where synergies with current businesses are most likely to exist.

Examples of growth strategy are listed in each quadrant of the Ansoff matrix on the right hand side of Fig. 8.4. Let us reiterate that the value of the Ansoff matrix is only to facilitate discussion and exploration of potential routes for growth within the specific circumstances of each case and each client. Some strategies might span across multiple quadrants in the Ansoff matrix, and Igor Ansoff himself noted that the simultaneous pursuit of penetration and related and unrelated diversification is essential to survival in the long term [153]. As options are narrowed down, developed, and ultimately implemented, many factors might influence both the success and scope of the chosen course of action.

8.3.8 Innovation Strategies

A special case of corporate growth may be made of innovation strategies, both for their peculiar return/risk profile and for the unusual approach that they may entail. Innovation is a key strategic lever for growth in most marketplaces for at least two reasons. First, innovation does not mandate the development of a new product or a new technology: all sorts of innovation, and in particular business model innovation [76], may lead to growth. Second, innovation does not mandate radical change. Most corporations today must deliver small incremental and ubiquitous innovations on a regular basis if they are to adapt, grow, and even just survive in their marketplace.

Let us first discuss the incremental type of innovation, which will be referred to as *Sustaining Innovation*. The more radical type of innovation, which lead to many interesting theories, will be discuss through the concept of *Disruptive Innovation* by Clayton Christensen [18] and *Blue Ocean* by Chan Kim and Renée Mauborgne [77], both of which were built upon many case studies spanning various historical periods and industries.

1. *Sustaining Innovation*

A sustaining innovation focuses on improving a given offering in a given market. It perpetuates the *current* dimensions of performance, as characterized by a reinforcement of interest in a pre-existing market for a pre-existing product/service line. This contrasts with disruptive innovations, which ultimately overtake existing markets.

In sustaining innovations, incumbents enjoy a clear competitive advantage compared to new entrants [18]. This is because they are highly motivated to win these battles with innovations that appeal to their most valuable customers, tend to be higher margin, fit their existing value chain, and match the "mental model" that they have built about how their industry works [18, 19, 75].

Finally, sustaining innovations have become necessary for the very survival of certain companies. For example in SaaS[2] companies, software updates are expected to happen regularly and are thereby bundled in the service that customers pay for [225]. In such circumstances, sustaining innovation becomes an integral part of the business model.

2. *Disruptive Innovation*

Disruptive innovation aims at creating new markets or else transforming existing ones. The first category addresses *non-consumption* while the second addresses (the needs of) *over-served* consumers [19].

Non-consumption defines a target market made of consumers who are currently not satisfied by available products. Where it becomes subtle is that consumers are not necessarily conscious that some of their problems could be solved in a better way, right now. So when the consultant attempts to define non-consumption, it is essential that he or she aims at addressing directly some problems to be solved, *jobs to be done* [19], rather than just inquiring about customer satisfaction for products already available in the target market. Consumers fulfill these needs by turn-arounds, non-optimal solutions, and thus opportunities abound in bringing to market a value proposition that better addresses these needs.

Moreover, by targeting unexplored market spaces, the competitive response from existing players is minimized because the innovation responds to the desire of a new market, one that does not qualify as the primary market for any existing player.

[2] SaaS = "Software-as-a-Service".

When the target of disruption is *over-served* consumers, a mainstream market exists that the disruptive innovator hits from the bottom of the performance level as perceived by current market players [19]. This is made possible because, to quote Harvard Professor Clayton Christensen, *technical improvements on the supply side tend to outpace technical needs on the demand side* [18]. In other words, customers become less and less willing to pay more for the added value.

From the perspective of the corporate organization, an over-served market is a mismatch between price and cost: while prices may easily be adjusted to customer willingness to pay, costs may not. As a result, the bottom line suffers. Incumbents in the mainstream market can hardly contemplate a strategy where to just go back to a former profit model that used to perform well in the past, because the marketplace has changed, new products have become benchmarks, and customers expect a lower price for older products. At this phase in the industry life cycle, customers often expect the added "unwanted" value for free [19], just because proxies matured.

So what are the incumbent's options? An incumbent in the mainstream market has evolved resources, processes and values that may be difficult to downgrade. Its model can thus more effectively serve high-end customers. High-end customers include two categories: buyers who, in contrast to overshot customers, do appreciate the recently added value (e.g. professional accountants still appreciate the enhanced functionalities of Microsoft software products even though households prefer the simpler Intuit's Quickbook versions), and buyers who have more emotional reasons to look for high-end niche products (e.g. sport car enthusiasts turn to Porsche's latest products rather than Subaru's). This is a key takeaway in Professor Christensen's thesis: disruption pushes incumbents toward high-end markets and creates circumstances favorable to new entrants [18].

From a tactical perspective, the need of *over-served* consumers in a marketplace is often most efficiently addressed by a new organization with an innovative profit formula [225]. The business model may then be re-invented to focus on new success factors such as simplicity, convenience, affordability, and accessibility [226]. This in no way implies that incumbents are doomed to fail to compete against disruptive startups since in fact, the new organization does not *have to* be a new entrant. It may simply spin off as a subsidiary from entrepreneurial initiatives within a large corporation. And with the right execution plan that fosters capabilities exchange between a large corporation and a new business unit [226], the two organizations may eventually leverage each other's strengths to an extent that makes it impossible for new entrants to compete. In conclusion, when a consultant works for a large corporate organization (an *incumbent*), building up an independent business unit is central to developing a disruptive innovation growth strategy.

How to uncover a potentially disruptive innovation? Again the answer has to be found in understanding customers and capitalizing on that understanding. To bolster innovation, a powerful approach to segment the market is what Clayton Christensen refers to as *Jobs-to-be-Done* [18, 19, 75]. The rationale is that the needs of individuals in a segment defined by demographic or psychographic variables such as age, income, or profession, can vary significantly [46]. Even specific needs as spelled out by actual customers is influenced and thereby restrained to the arenas of pre-existing offerings. Thus Clayton Christensen proposes to directly address these needs by articulating what customers are trying to achieve and then use this circumstance-based metric as the basis for market segmentation. In other words, the concept of Jobs-to-be-Done proposes to segment a market out of the answers obtained when asking the following question, or some refined set thereof: what jobs are customers trying to get done? [46].

Innovating, and in particular disruptively, does not necessarily require new technologies, nor even a new product. Innovative business models (how a company creates, captures and delivers value) can disrupt markets as severely as any new technology [76]. It suffices to look at companies such as Netflix, Google and Amazon to appreciate how business model innovation without any new technological breakthrough may redefine the dynamics of entire industries.

3. *Blue (vs. Red) Ocean*

The concept of Blue Ocean (2004 [77]) is very close to the concept of disruptive innovation and was developed only a few years after disruptive innovation (1997 [227]). Blue Ocean and Disruptive Innovation primarily differ in the details of how its founders defined the application circumstances, i.e. in which situations does the proposed strategy most effectively apply. In particular, Blue Ocean is closer to disruptive innovations that have potential to create *new* markets rather than the sort of innovations that disrupt *existing* markets. In fact, the name Blue Ocean originates from the metaphoric red sea that characterizes a highly competitive saturated market, and thus exclusively refers to innovations that compete against *non-consumption*. But again, strategists beware. These definitions are theoretical conventions: new markets sooner or later affect existing markets. This is just a matter of time. Anticipating these future impacts is the very job of a strategist. In this sense, it may be useful for a consultant to consider Disruptive Innovation as the most comprehensive framework because it better emphasizes the potential of disruption *between* different marketplaces.

Blue Ocean and Disruptive Innovation also differ in the recommended course of action to innovate. Blue Ocean here has undeniable value: it brings original perspectives on how to handle markets waiting for creative innovators. Blue Ocean builds upon the concept of *Value Innovation* (1997 [227]) from the same authors, which one posits that in absence of competition, *differentiation* and *low-cost* strategies may be leveraged both at the same time. This represents a more profound proposition than a simple short-term tactic because it addresses a key weakness of a central dogma in strategy –the specialization dilemma that every company faces between differentiation and low-cost. It proposes that low-cost can in fact be achieved through perpetual search for "innovative" differentiation. In other words innovative differentiation opens the door to unexplored market spaces (the blue oceans) in which low-cost may be achieved before competition sets in, and dynamically navigating this opportunity space enables a company to sustain low cost advantages.

As for the different types of portfolio matrices introduced earlier, Blue Ocean and Disruptive Innovation are two different approaches each with its own set of advantages and application areas; they might thus be combined to develop custom made strategies that best fit a given case and a given client. Common grounds between the two approaches are the need to differentiate when disruptive and blue ocean innovation apply versus when more conventional growth strategies apply (sustaining innovations), and also the need for creating new customer segmentation schemes to address over-consumption and red ocean battles in ways that are not or cannot be leveraged by incumbents.

8.3.9 Signaling

Signaling is a form of communication between competitive organizations. Antitrust laws restrain direct forms of communication because these could open the door to unfair market domination by just a few players, and in turn not benefit consumers.

As in poker, signaling your intention is a key lever for all players, and may entail very subtle strategies. And as in poker, it may result in mutual benefit for all players. For example, if a market player publicly acknowledges a certain budget for *advertising* that matches the investment made by its competitors, it signals a potential agreement for working with a certain level of cost-effectiveness in advertising. Indeed if market players were to keep increasing their respective budget in response to each other's latest campaign, it would most likely decrease cost-effectiveness without significantly affecting market shares, and in turn be detrimental for all players. The market opportunities are only as big as the market.

Signaling is often used to avoid *price wars*, which would be detrimental for all players [228]. For example, if company A publicly announces that a price cut is only for a limited time, it signals to its competitors (company B) that this is not a long term intention to keep prices low. In this example, company B may either adjust its own prices temporarily, or wait for a later opportunity to cut prices. If it waits, it may expect that company A will return the favor and not adjust its prices right at the same time, for the sake of sustained cooperation. If company A did so, it would signal disregard for company B's recent courtesy, a "winning-only" attitude, and thereby engage in a long term battle of price cuts.

Take this other example: a company commits to a low-cost business model and publicly announce that it will meet or beat any competitor's price. In this case, it is signaling a decision to commit to a *low-cost model*. Only competitors who have the potential and intention to compete at the bottom of the market/performance level may then want to adjust. Others, such as relatively small companies, may instead explicitly retire from the low-end section of the market and sharpen their focus on higher-end niche opportunities. Again in this example of *long term* price cut signaling, there is potential for mutual benefit between all market players as well as consumers.

One subtlety of signaling is that different market players might not interpret messages in the same way, or literally leverage it against each other and in particular

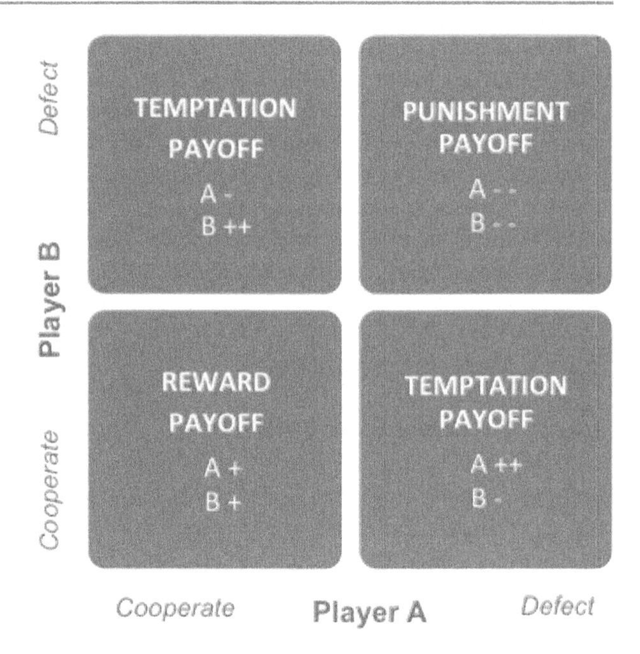

Fig. 8.5 The Prisoner's Dilemma payoff matrix

against the company that signaled information. In the example above where a company reacts in a courteous manner and may expect reciprocal favors in the near future, a competitor who would disregard the courtesy and engage in a battle of price cuts would benefit from the advantage of having started the campaign earlier. Signaling often leads to this kind of dilemma, referred to as the prisoner dilemma [229].

The concept of prisoner dilemma is rooted in game theory [230], and was formalized in the 50's by the mathematician Albert Tucker for a number of applications in economics, sociology, biology and psychology [231]. It addresses a decision-making dilemma between pursuing either self-interest or cooperation, whereby two organizations almost always enjoy a better outcome if they both collaborate than if they only pursue their own self-interest. But here is the dilemma: if company A decides to cooperate and company B decides to pursue its own self-interest at the expense of company A, then company B enjoys the best possible outcome. Thus, in practice, keys to handle the dilemma are trust, self-confidence, perception of each other's confidence, fear of retaliation. Basically all attributes of a poker game…

To illustrate the dilemma, let us think of Coca-Cola deciding to cut prices. PepsiCo may have no choice but to follow, which in turn is likely to maintain the former equilibrium in term of market shares between the two companies, and result in a significant drop in profits for both companies. A price drop breaks an implicit agreement between Coca-Cola and PepsiCo to keep prices high and maximize profits. If Coca-Cola drops its price but PepsiCo continues to keep prices high, Coca-Cola is "defecting" while PepsiCo is "cooperating", and Coca-Cola wins. Only if Coca-Cola and PepsiCo both stick to the spirit of the implicit agreement to keep prices high may they both maintain a relatively high margin.

The different scenarios may be further developed, quantified and organized in a *payoff matrix* (illustrated in Fig. 8.5). The essential take away of the prisoner dilemma is this: both players benefit from cooperating as long as no one breaks the pact. This pact cannot be explicit (by law!) and is often ill defined given that short term tactics such as price cuts and aggressive advertising campaigns are classic textbook marketing weapons. It is natural thus that, in any given marketplace, signaling takes an increasingly concrete form (via public relations, TV broadcasts, etc) as companies get larger, older, and more familiar with each other.

8.4 Marketing Strategies

8.4.1 Customer Segmentation

Defining customer segments serves many purposes, such as identifying the needs and economics of each segment, identifying the players serving each segment, and effectively aligning a value proposition. There are always many dimensions along which to segment a market. The interplay between these many possible dimensions makes segmentation an art, i.e. there is no single right answer to segmenting a market.

New information technologies and big data now enable detailed tracking and analysis of transactions [13] which, in combination with more flexible value chains, enable companies to tailor their communications and products and serve segments as small as a single consumer [232]. Nevertheless, segmenting customers into "clusters" remains necessary. This helps design more effective marketing programs and sustain economy of scale. Customization, it may be argue, is not as much a new trend in segmentation strategies as it is a new characteristic of particular segments – a modern customer desire that may be fulfilled using big data analytics [233].

Dimensions along which to segment a market may be categorized into four groups [46]:

- **Geographic**: Country, state, region, city
- **Demographic**: Age, sex, income, marital status, education, profession, ethnicity, religion
- **Psychographic**: Lifestyle, activities, interests, opinions, personality (e.g. risk taking, status seeking, conservative, compulsive)
- **Behavioral**: Benefits sought, purchase behavior (e.g. regular purchases, gift, vacation, seasonal, responsiveness to price cuts and promotions), brand loyalty

And the criteria used to evaluate the attractiveness of these dimensions, and the segments that result from working along these dimensions, may be categorized into four groups too [46]:

- **Measurability**: How well may the segments be identified and their size measured

- **Accessibility**: How well may the segments be reached (advertisement, distribution)
- **Profitability**: How profitable might the segments be, and how did they grow recently
- **Compatibility**: How well do the segments fit the organization's capabilities and competitive strengths

Despite a common tendency in management consulting to segment market through relatively easily quantifiable geographic, demographic, or psychographic attributes, many researchers suggest that behavioral circumstances, and in particular explicitly articulated needs and benefits sought (*jobs-to-be-done*), should always be the ultimate dimension along which to segment a market [19, 46, 75]. This is because all other types of segmentation schemes may contain large degrees of variability *within* each segment regarding actual needs and benefits sought. Identifying what customers are trying to achieve when purchasing a product is ultimately what a segmentation strategy intends to do, and innovating a new segmentation strategy can be as powerful as innovating a new product or a new business model [19].

8.4.2 Market Analysis

Market analysis is similar to customer segmentation but instead of looking at individuals, it identifies holistic attributes such as total potential market size, life cycle, basis of competition, role of government, barriers to entry, cost-structure benchmarks [100]. A market as defined here may be an entire sector, or just one segment.

The *product life cycle* is a commonly used concept because it adopts a similarly shaped curve across all industries [234]; only the timescale may change. The life cycle is defined by four phases: introduction (small slope), growth (exponential slope), maturity (null slope), and decline (negative slope).

Specific marketing approaches apply for specific phases of a product life cycle [100]. For example, the introduction of a new product where purchases are made by the so-called *early adopters* may require particularly high investments in advertising and education. It is a time when both features and prices can be adapted to better position the product with respect to unanticipated customer needs and present/future competition.

During the growth phase when purchases are made by the so-called *early majority*, competition has generally intensified and customers have the choice between alternative products. In this phase a more balanced marketing approach should be adopted between optimal design (quality/features), production efficiency and advertising efficiency [100].

When the life cycle has reached maturity (*late majority*), the market has the characteristic of *over-served consumers* [18, 75] described earlier in this chapter. Features that most customers really want have become standardized and the introduction of new features essentially increases cost but not profit. Mainstream customers are over-served, price competition intensifies, low cost and mass-market distribution pervade. Here brand loyalty and advertising become keys to success [100]. Market players who react by migrating farther toward more differentiated or higher end segments have no choice but to contend with smaller niches [18].

Finally, a declining phase is characterized by a market-wide decreasing profitability, with either noticeable decline in prices or in number of sales/market players. New customers are referred to as *laggards*. Advertising is relatively inefficient at this point because customers are familiar with product standards and do not value new features. The marketing approach with highest potential often revolves around *trade relations*, i.e. pre-established networks of business partners and distributors [100].

Sustaining a business requires leveraging the product life cycle by identifying and taking full advantage of the growth and maturity phases, and getting ready to face decline when the time comes. For example, mature cash-generating products may fund the development of younger and potentially successful products. Growth and innovation strategies should be developed along appropriate time horizons, which may vary from just a few months (e.g. clothing apparel sector), to a few decades (e.g. certain energy sub-sectors).

In contrast to relatively generic marketing strategies that accompany a market life cycle, the *basis of competition* and *key success factors* may be very specific to each market. For example, in heavily regulated markets such as healthcare, understanding government policies and their implications may provide a powerful competitive edge. It enables executive decisions to overcome regulatory hurdles and to reach stakeholders (e.g. physicians, hospitals, insurances, patients) more effectively than competitors. In other markets, for example in the news industry, this factor becomes close to irrelevant.

Examples of key success factors include economy of scale, brand recognition, reputation, customer loyalty, R&D capabilities, access to highly skilled workforce, optimum capacity utilization, ability to adopt new technologies, proximity to key markets, control of distribution channels. This is really just to name a few: IBISWorld identified a total of 250 key success factors across its platform of 700 US market analysis reports [235].

It is important to remember that, as discussed earlier, the definition of a market and segments thereof is not static. The basis of competition and key success factors may slowly evolve or be radically disrupted. An innovative segmentation scheme for example may quickly redefine the boundaries of a given market segment and thereby its key success factors. The strategies of Disruptive Innovation and Blue Ocean described earlier rest on that premise.

8.4.3 Competitive Analysis

The above concepts of customer segmentation and market analysis help the consultant better understand the market segments in which a company aims to focus. For a marketing strategy to be fully effective, the consultant also must benchmark similar products and substitutes, and overall gather intelligence on the competitive landscape.

Analyzing the competitive landscape requires addressing questions and potential success factors concerning: core competencies (activities that a company does better than others), resources (workforce, technology, R&D, patents, cash, sales force), cultures (short- and long-term objective, specialization, focus), market fragmentation (key players, respective shares), positioning (price, promotion, distribution, packaging), barriers to entry, historical performance.

The *SWOT analysis* [236–238] is a popular framework that relates the strengths (S) and weaknesses (W) of a company with the opportunities (O) and threats (T) brought about by the environment (competitors, customers, government, etc). One constructs a SWOT table by representing a list of potential success factors in one dimension of the table and listing four entries in the other dimension: strengths, weaknesses, opportunities and threats. In order to better understand the threats and opportunities from the perspective of different market players, it can be useful for the consultant to build a SWOT matrix both for the client and for the closest competitors.

8.4.4 Positioning

"The perception of your product in the minds of consumers is more important that the physical reality of the product's attributes. What consumers believe is their reality"
Al Ries & Jack Trout, 1972 [239]

Positioning may be defined as the way by which a company attempts to create a distinct impression in the customer's mind. It addresses the place that a product, brand, or group of products occupies in the consumers' minds relative to competing products [239]. The positioning of a product is affected by the entire company's *public image*, i.e. the opinion that current and potential customers have of the company [240]. The consultant should thus evaluate how consistent this public image is with the image that the company has of itself.

In this sub-section, two simple frameworks for developing positioning strategies are introduced. These represent useful *checklists*, mental roadmaps for thinking about marketing and action planning. A more advanced discussion on marketing strategies may be found in the next chapter.

The 4 P's marketing mix

The 4 P's framework is no strategy; it is a convenient way to describe the menu of options that are available to market a product. To the consultant, it can be a convenient roadmap for action planning [148]. Each option may be weighted against pros and cons that reflect the specific circumstances of a given case and a given client.

The two following guidelines should be followed when using the 4 P's. First, discussing the strategic value of a given option should always be done relative to the specific circumstances of the case and the client. Second, the 4 P's are coupled to one another; so each component should be designed to mutually support each other's objectives.

The Product

Many characteristics of the product may be leveraged for positioning, e.g. features (functions, size, color, shape, style), fit with the rest of the product line and business units (synergies), reliability (warranties, return policies), convenience, packaging (color, size, shape, protection), services (punctuality, courtesy), branding (perceptual map, labeling style, co-branding arrangements)

The Places

Distribution and sales channels may be exclusive (sales in one exclusive outlet), selective (sales in a few select outlets), or intensive (mass distribution in as many outlets as possible). The best option depends on the nature of the product and on corporate objectives (level of control desired, target margins)

The Price

Pricing a product should take into account costs, competitors, and consumers (perceived value and demand elasticity), as predicated by standard pricing frameworks (see chapter 5). Additional options include *dynamic pricing* (e.g. seasonal demand fluctuation) which is famously leveraged by airline companies, *skimming* (high price upon product introduction to maximize profit before competition kicks in), and *penetration* (low price upon product introduction to maximize market share before competition kicks in)

The Promotions

Promotions include all advertising and sales initiatives. Promotion strategies may be categorized as *push* or *pull* strategies depending on whether the goal is to pull buyers into the outlets that carry the product, or to push sellers (sales channels) to sell more. Or both. Following Steven Silbiger in [100], let us differentiate six types of promotional campaigns:

- **Sales Promotions** – Coupons, rebates/refunds, samples, premiums (free products offered upon purchase of other goods), games/contests. The game *Monopoly* that re-appears every year at McDonald's is a good example of sales promotion
- **Channel Promotions** – In-store demonstrations, point-of-purchase displays (e.g. *end caps* at the end of the aisles in CVS and Walgreens stores) and sales incentives (e.g. slotting fees, contests rewarding best sellers, free merchandises). Trade-shows are also leveraged where companies meet retailers and wholesalers to cultivate a network of distributors
- **Advertising** – Options may vary from traditional newspapers, magazines and TV advertisements to modern website banners, pop-ups, and keyword Internet searches
- **Personal Sales** – Sales personnel and networks are key to marketing a product in certain industries. A pharmaceutical company for example often needs to obtain doctors' buy-in based on complex lab results that have various degrees of reliability. These results are most efficiently communicated with 1-on-1 meetings, symposiums, visio-conferences, etc.

- **Direct Sales** – Internet outlets are used together with emails and catalogs. Beyond Amazon, niche outlets may be used as it may affect consumers' perception of the product
- **Public Relations** – Initiatives range from appearance in press conferences, movies and news (i.e. publicity) to *viral* marketing (buzzs, tweets), and sponsorship of prestigious or charitable events

The 5 C's marketing mix
The 5 C's framework [241] is less action-oriented than the 4 P's but has more value in the breadth of factors that it radiates through. Indeed it encompasses the 4 P's and addresses all players and environments that may impact the business and thus the positioning of a company.

The Company
This includes core competencies, resources, processes, objectives, values (culture) …and 4 P's

The Competitors
The competition may be direct (similar offerings), indirect (substitutes), or potential (anticipated in the near-future). Again it should cover core competencies, resources, processes, objectives, values, 4 P's, but also market shares, historical performances, and forecasts

The Customers
Details are discussed in the sections *customer segmentation* and *market analysis*: it includes segment characteristics, size, growth, needs, motivations, behaviors, sales channels, etc.

The Collaborators
This includes suppliers, distributors, business partners sharing assets and liabilities, but also niche experts and agencies that specialize in HR, IT, law, international trade, etc.

The Climate
This is defined by macro-environmental factors such as politics (government's role and policies), economic (inflation rate, exchange rate), social (societal culture, norms, education levels) and technologies. Analyzing these four factors is referred to as a *PEST* analysis [242].

8.4.5 Benchmarking

Many factors influence the success of even the best thought through strategies, including organizational design, use of technology, staffing structure, ability to network. What ultimately brings the execution of a strategy to success in a given marketplace is the result of a combined effect of many key factors together with the pace at which the company innovates along these factors. Thus, studying how world-class companies operate may accelerate the thinking process when developing a strategy, and help identify key success factors [243]. Benchmarking world-class companies do not prevent differentiation because the consultant can always innovate with new cock-tails of best practices.

Management consultants often utilize *benchmarking* in order to quickly get familiar with the potential strategic levers that characterize a marketplace, based on what the best performing companies share in common. Products, processes, competencies and cultures, can all be benchmarked [243]. The key to successful benchmarking is to understand not only the performance results of benchmarked companies but also what lies behind these results [5]. Only then will the consultant grasp how productively the benchmarked factors may benefit the particular circumstances of his/her client.

A recipe for benchmarking may consist of the following steps (Adapted from Ref. [5]):

1. Select key performance figures to benchmark
2. Identify *best-performer* companies along the chosen figures
3. Measure the performance gap between client and benchmarked companies (trends/forecasts)
4. Identify factors that account for the difference in performance
5. Articulate how the client performs along the identified factors (client's practices)
6. Articulate how best performers perform along the identified factors (best practices). A consultant can more easily survey competitors that the client. He/she may even visit in person the best performers to develop an understanding of best practices
7. Determine short- and long-term performance goals
8. Prepare plans and schedules for implementing best practices in the client's organization
9. Set up performance measure systems
10. Assist in the implementation

The following is a sample of recurrent themes in benchmarking as observed across different marketplaces today [5]:

- Most successful companies excel in customer focused practices
- They have developed an exceptional ability to learn and an open learning culture (e.g. *Agile* method)
- They invest effectively in intangible assets and particularly in human capital
- They tend to use some version of a *lean* production system (*i.e.* a continuously optimized tradeoff between cost and value-add for every activity)

9 Principles of Strategy: Advanced

In this final chapter, on the topic of *advanced* strategies, a selection of topics is discussed through essays across the functional (Sect. 9.1), business (Sect. 9.2) and corporate (Sect. 9.3) strategies. As noted earlier, the merit of defining these three categories is purely pedagogic: developing a strategy should always involve holistic approaches that appreciate the many potential interactions and intricacies within and beyond the proposed set of activities.

9.1 Functional Strategy

9.1.1 Performance Strategies

Performance (a.k.a. *productivity*) may be defined as a measure of the quantity and quality of what is being produced in relation to the resources used in a given macroeconomic, institutional, social and natural environment [244].

This section discusses the scope of performance measures and why too narrow definitions are misleading. Narrow definitions include the physical output/input ratio, labor productivity and capital productivity. Performance indeed is the result of a combined effort of many stakeholders, each with different objectives and perceptions about organizational effectiveness, and thus a measure of overall company productivity shall consider a dynamically evolving set of internal and external factors. With this understanding, the discussion moves to commonly used approaches to improve performance. One special type of performance improvements program, Total Quality Management [245], is described in detail in the next section.

How to measure performance?

Productivity is traditionally perceived as a ratio of [physical] output to input. This measure is misleading because the measure of input is ill-defined: it may vary with such factors as labor, capital, technology, processes, systems, knowledge, skills, management, cultures, logistics and others. The output itself may also vary with

J. D. Curuksu, *Data Driven*, Management for Professionals,
https://doi.org/10.1007/978-3-319-70229-2_9

customer satisfaction, customer perception, shareholder perception and product life cycles. Many such factors are not under the control of the client. For example, the net value of certain resources such as skilled labor, land and material, may easily change upon new governmental policies, international trade agreements, technological breakthroughs, or social climate.

Profitability and return-on-investment are more useful indicators of overall company performance because they implicitly take into account many factors and focus on a key driver of success: the short-term financial performance. But they lack obvious dimensions that are prerequisites to sustaining a long-term success, such as customer-based measures (satisfaction, loyalty, market share), resource-based measures (skill development, employee retention), and process-based measures (operation excellence).

Benchmark studies [5] show that top companies acknowledge the need to create performance measures that focus on at least four drivers: financial performance, shareholder value, employee satisfaction and customer satisfaction. The consultant should draw together a set of measures by taking the perspectives of various stakeholders. Gathering information through surveys, observations and direct interviews of customers, employees, shareholders, owners, suppliers, communities, in turn, will prevent the consultant from the shortsightedness that often results from focusing on a single measure of success. As mentioned earlier, performance is the result of a combined effort of many stakeholders with different objectives and perceptions...

Innovation is required for long-term performance

Many factors such as social climate, consumer habits, new technologies, cannot be controlled nor predicted. In consequence, companies should always be willing to take some [reasonable amount of] risk and foster an innovation learning process. Most successful companies today have shifted their organizational priorities from a traditional *input-focus* and *resource allocation* to a modern concept of *customer-focus* and *resource attraction*. In other words, they foster entrepreneurial spirits and innovation-friendly learning environments that focus all attention on the value added for customers. The *Agile* method for example, emphasizes project delegation to self-organized "squads" of 4–5 people and constant iterations across the network of stakeholders. The traditional rating schemes of employee performance are being replaced by more frequent employee-manager conversations and coaching because these better encourage collaboration and risk-taking. They diminish workers' anxiety and deliver better focus on their actual performance rather than on managers' judgment [246].

Other innovative practices for performance improvement that have become commonplace in today's markets are *Information Technology* (just-in-time delivery, reduced transaction-costs, intelligence on customer purchases), internal *knowledge sharing* (weekly round-table discussions, communities of practice) and external *symbiotic partnerships* (one-stop-shop service offerings, coexistence between competitors).

These innovative practices are prerequisites for staying competitive in today's high tech markets. But the consultant needs also study the *fit* of these resources and priorities with the specifics of the client's strategies and processes [19]. The integration of innovative ideas within the client's capabilities is what ultimately unleashes, or in contrast prohibits, the full potential of these ideas.

Strategies to improve performance

As a first step, the consultant should assess whether small continuous improvements that require little investment may achieve growth objectives or if more radical interventions are needed.

Incremental growth in productivity may be achieved through reducing the generation of waste and unnecessary capital expenditure. For example, a *preventive* waste management system may be developed by reviewing technologies, process designs, work methods, choice of materials, inventories, machine maintenance, space utilization, employee's problems, and each time focusing on what really provides value to the customer.

A more *radical* strategic intervention may be time consuming, involve large expenditures/risks and be met with much resistance. But it also holds the key to higher payoffs and thus in certain circumstances it is the only viable solution.

Once the scope of the strategic intervention (incremental vs. radical) has been articulated, the consultant disposes of a palette of *performance improvements programs* that includes hundreds of tools going from simple brainstorming to benchmarking to more sophisticated IT-enabled one-to-one marketing techniques.

A performance improvement program should fit the specific circumstances of a given project and a given client. The consultant should thus first *diagnose* current results and challenges through extensive monitoring. Then, the cost of investment and time horizons should be compared with the likely benefits and risks. At this juncture, the organizational readiness and consultant's awareness on productivity issues as well as his/her own competence in using particular performance improvement programs, should be considered. The consultant should make sure that top management will commit and all relevant personnel understand the program. He/she should inquire the degree of internal resistance to organizational change and the degree to which his/her team may experiment with new solutions [5]. Ideally, the program will have universal appeal, cross-over capabilities, and create a common language for diverse professional groups.

The approach to implement a performance improvement program may be structured in five phases:

1. Identify the situation and opportunities for potential savings. For example, the consultant may identify products that represent the largest output and look into their cost elements
2. Select a program (Table 9.1) based on potential for presenting the largest payoffs-to-risk ratio
3. Obtain support from all parties in the client organization that may affect the successful completion of the designed method

Table 9.1 Thirty examples of common performance improvement programs

Sample of performance improvement programs		
Simple	Intermediary	Complex
• Brainstorming	• Benchmarking	• Just-in-time
• Gantt charts	• Job enlargement	• Mission and vision statements
• Productivity training	• Value analysis	• Supply chain analysis
• Process control	• Cost control	• Knowledge management
• Activity analysis	• Conflict management	• Strategic alliances
• Waste reduction	• Cycle-time reduction	• Learning organization
• Work organization	• Cross-functional teams	• Customer segmentation
• Work simplification	• Virtual teams	• Total quality management
• Agile teams	• Six sigma/Lean	• System thinking
• Balanced score cards	• Strategic alliances	
• Design thinking		

4. Articulate specific tasks to implement the selected program
5. Implement performance measures and control systems to evaluate improvement achieved once the selected program has become functional

9.1.2 Quality Management

Total Quality Management (TQM [245, 247]) is an approach for improving performance that has grown over a period spanning 40+ years. It was initiated in Japan by popular optimization and quality control movements. TQM illustrates a universal trend across global organizations and marketplaces from a focus on product quality to a focus on delivering value to the customer. TQM envisions a definition of *quality* in the utmost holistic perspective that an organization is willing to consider, involving all level of strategic planning, management control, process design, operations and an emphasis on customer satisfaction. It is a continuous process that has more to do with long-term cultural changes than any particular program implementation. This section discusses some principles of TQM and ends with a general 5-step approach to implement a TQM.

Quality is a relative term. It may be defined as "conformance to requirements". A special case and modern illustration of quality is the *ISO* norm, a series of international standards for quality systems recognized and adopted by 150+ countries [248]. An ISO 9000 certification [249] for example involves a third-party audit and has become mandatory for certain public sectors and medical contracts. Beside ISO norms however, there is no preset requirement for TQM, which may be adapted to all sorts of needs and circumstances [247]. To guide consultants in developing a TQM strategy for their client, Milan Kubr [5] provides the following definition:

> *"TQM advocates the development of true customer orientation, teamwork and inter-unit cooperation, structured problem solving, a reliance on quality assurance standards and measurement, a system of rewards and recognition for excellence, and long term commitment to the ongoing process of improving quality. It creates an environment that contributes*

to positive morale and recognizes that products and services embody the efforts, creativity, values and collective personality of their producers. The fundamental engines for TQM are empowering, energizing and enabling."
Milan Kubr, 1977

TQM requires result-oriented strategic policies that members of the organization can understand and follow, and some empowerment systems that will encourage them to commit to TQM. Simple examples include trainings, reward systems, TQM-dedicated investment and management teams. Quality control assessment systems should also be deployed to sustain a TQM environment in the long-term. Some basic principles associated with TQM are:

- Quality measurement shall be customer driven
- Objectives shall go beyond meeting customer satisfaction and upgrade to customer *delight*
- Top management shall strongly commit because of the holistic nature of TQM
- Employees shall be the key "levers" for improving quality
- Teamwork shall be emphasized because all activities in the value chain have potential for quality improvement

The approach to implement a TQM strategy may be structured in five phases (Adapted from Ref. [5]):

1. **Diagnosis I: Identify high potential improvement areas**
   Quality needs be defined as conformed to specification (internal, benchmarked, or ISO-like) and measured through control systems that enable the financial consequences of quality problems to be quantified
2. **Diagnosis II: Apply the "total" concept**
   Focus should start with the most easily assessed resources, processes and decision-making systems, and then move toward management and customer issues
3. **Diagnosis III: Marshall the capabilities of the organization to delight customers**
   TQM aims to fulfill not only manifest needs (i.e. what customers say they want), but also latent needs which can be defined as what the client's customers would ultimately want if they were aware of and could leverage the full potential of the client's capabilities. Studying latent needs, jobs "yet-to-be-done" in the customer's mind, may open the door to unexpected high potential innovations for quality improvement. This aspect is what will ultimately unleash the full potential of an agile, strategic transition toward TQM
4. **Implementation**
   Commitment from top management should be sought out as early as possible. Based on the diagnostic results, specific quality improvement programs should be defined, then responsibilities and process ownership across all teams in the organization should be articulated. Communication flows between teams, departments, and with suppliers and customers, should all be optimized to elimi-

nate processes that do not add value for customers. Simple examples include bottlenecks that prevent timely delivery or output requirements. The implementation phase should include tailored training exercises across all hierarchical and functional levels in the organization to help develop agile and total quality mindsets and skillsets

5. **Support and review systems**
   A sustainable TQM strategy requires adequate support systems (both financial and non-financial) from top management across the organization, and quality measurement systems that will enable top management to regularly assess how much success is achieved and decide whether to maintain, modify, or eliminate diverse TQM initiatives

9.1.3 Operation Strategies

Operations materialize the entire value chain. Related consulting activities are often referred as *supply chain management* (for a precise distinction between value chain and supply chain, see Chap. 3). The value chain operations include procurement, manufacturing, logistics, distribution, sales and marketing. A key feature of operation strategies is the interdependence between activities in the value chain: often the problem diagnosed by the client before the consultant was enrolled is a sign of much broader needs in the organization and its value chain [5].

Operation strategies may be seen as just a special case of the performance improvement strategies introduced above. As such they require choosing between incremental or radical initiatives, and establishing performance criteria. Examples of performance criteria for operation strategies include:

- **Speed** (processes, development cycles, capital rotation)
- **Quality** (manufacturing, value-added focus, cross-functionality, pace of innovation)
- **Customer focus** (customization, streamlining, competitive offerings)

Beside the general approaches described in the two previous sections, it is useful for the consultant to familiarize with program-focuses that more specifically pertain to working on operation cases, as proposed by Hayes, Wheelwright and Clark [250]. Focus indeed could for example be on products and services, management, systems, schedules, technology, vertical integration, human resources, facilities or distribution channels.

Given the diversity of available options for improving performance (Table 9.1), the consultant should be careful not to generate too many projects, or projects that cannot be absorbed by production facilities and staff in the client organization. Benchmarking competitors can often help understand and optimize value chain characteristics and in particular their potentially complex interdependencies.

9.1.4 Information Technology Strategies

As described in Chap. 2, the recent emergence of computerized technologies and the so-called revolution in the economics of information provided a disruptive growth engine for the management consulting industry as a whole [4]. It gave birth to a regimen of diverse and specialized IT consulting firms [12]. And by affecting virtually all industries, it naturally transformed the profession, bringing in an entirely new wave of client problems and strategic opportunities [13, 251]. This section discusses key trends in IT-enabled business strategies and then presents some common types of consulting assignments centered on assisting a client with IT strategies.

What drivers of change surfaced with the growth of IT?
Impactful drivers of change relate to increased standardization, accessibility, availability, mobility, and access to large data centers. These data centers provide almost unlimited storage of *big data* on highly secured machines that can be accessed almost instantaneously from anywhere. In short, IT has enabled businesses with end-to-end flow of information across entire value chains, from suppliers to customers.

The revolution in IT has shifted challenges from "information" to "knowledge"
The abundance of data that can be obtained for free or at low cost has created a new challenge of information overload, which in turn fostered theoretical and practical development in tools and methods capable of finding useful information in noisy data coming from multiple sources [52, 97]. As described in Chaps. 2, 6, and 7, entirely new fields have emerged in the domain of artificial intelligence and machine learning [66, 96, 252]. These technical skills are becoming essential in business areas that used to require exclusively "soft" skills [98]. For example, big data analytics tools enable marketing departments to draw detailed individual profiles of thousands of customers almost instantaneously based on their credit card purchases [52, 97]. In a few days, they can produce a quantity of data so enormous that for many purposes, only sophisticated statistical methods have a chance to uncover any significant pattern in the data [66, 252].

IT has created new market spaces
The increased demand for skilled workforce that can implement, maintain and effectively use large scale information systems has created many opportunities for external IT organizations [69]. Although perpetual innovation in IT is required (see discussion below) and outsourcing limits innovation abilities [253], it is not economically beneficial for most companies to maintain a permanent workforce to develop and insource their own IT system [69].

Opportunities have floundered for management consultants who are now expected to bring highly diverse skill sets and bridge the capability gap between those who understand the technology and those who understand the business need.

This type of hybrid skillset has become a key competitive advantage in the management consulting industry as discussed in Chap. 3, and will become an even more important lever in the future (see Ref. [4] or forecasts deduced from scenario analysis in Sect. 2.3). This is the reason why the book on management consulting that you are currently reading contains entire chapters dedicated to quantitative data analysis (Chaps. 6 and 7). In today high tech environment, a management consultant without this skillset is akin to a librarian without a computer. There is no need for every librarian in a library to possess a computer, and a general knowledge of what a computer may do rather than the technical details of how it does it is certainly sufficient. But in today's digital age, a library without personnel capable of handling a computer cannot sustain its business.

IT has brought perpetual innovation into the spotlight
New information technologies enable innovative competitive differentiation by changing the way a business connects to its suppliers, customers, and partners [99]. Some IT strategies focus on processing market feedbacks to regularly innovate and test new products, some focus on market development using Internet as a vehicle for expansion, some focus on re-engineering the value chain for increased product quality or reduced cost. The potential of IT strategies is growing everyday as incumbents and new players innovate and experiment. A special emphasis should thus be put on developing a portfolio of IT innovative approaches, taking (reasonable amounts of) risks and fostering an IT learning process. Innovation is required in the quest for competitive IT-enabled strategies.

IT has created new management consulting activities
The following presents some of the most frequent types of assignments that consultants may be brought in to assist their client with IT strategies (adapted from Ref. [5]):

- Evaluate quality of IT systems, which translates into customer *and* employee satisfaction (because employees are the end-users of the IT system). Flexibility to meet unforeseen needs in the future is essential to measure IT quality
- Benchmark IT management practices and processes
- Develop new IT systems (initiated by internal incentive or in response to competitors)
- Assist with IT project planning and management
- Re-engineer database design, workload distribution, process optimization
- Develop an IT-savvy culture in the organization via training programs
- Review certain business strategies in the light of recent IT developments
- Compare software systems and vendors
- Evaluate outsourcing versus insourcing IT capabilities

9.1.5 Turnaround Strategies

A turnaround assignment calls for emergency measures that have to be evaluated, chosen, and executed in an unusually short amount of time. A characteristic challenge from the perspective of the consultant is a sense of panic and emergency in the client organization [5], including executives who are under unusual pressure from shareholders. The first step in the consultant's intervention is to perform a quick preliminary diagnosis to evaluate the overall situation from his/her own perspective, and assess how much potential for a solution there might be. The consultant might uncover that any potential turnaround operation will involve prohibitive costs. If the situation appears unsolvable the consultant might have to back off... when liquidating a company is the best option, no amount of consulting will avoid the ultimate fate of the company.

According to Milan Kubr [5], *poorly performing companies are often a hotchpotch of ill-assorted businesses*, the result of previous acquisitions, and what is then needed in more clarity of focus and a return to core strengths. When engaging into a turnaround case he advises the consultant to:

- Maintain a dialogue with creditors and prioritize payments that cannot be postponed, to avoid alienating external bodies that could initiate legal actions and end the game prematurely
- Focus on quick wins that will produce immediate savings or stop further deterioration (recruitment freeze, restriction on foreign travels, increase work discipline). At least it should help create an atmosphere where everyone involved realizes the gravity of the situation
- Invite employees to contribute to the turnaround in every possible way
- Identify both internal and external causes of the problem and look for aspects of the business with solid and sufficient potential to turn around the situation

A turnaround plan may be structured in five phases:

1. Identify causes of troubles: The consultant should overview a large number of internal factors (e.g. management issues, internal conflicts, unions, lack of competitive technology/process) and external factors (e.g. industry-wide depression, prices of supply, governmental factors)
2. Set objectives: The consultant should focus on what most affects cash flows. For example: increase sales, increase capacity utilization, increment profit expectation, finance growth only from profits
3. Articulate a strategy and quick wins: The fastest way to reach turnaround objectives is often to terminate some activities, cut costs, reduce payroll, downsize, change processes, merge/acquire
4. Implement changes and execute quick wins
5. Monitor the plan to evaluate progress and enable the client to make adjustments

9.1.6 Downsizing Strategies

Downsizing is a special case of turnaround. It consists in the reduction of the workforce and deserves a special note because it presents high potential for deteriorating an organization. The negative impact of downsizing on productivity, morale and loyalty, is popular. The hidden economic and social costs of downsizing are less popular, but even more catastrophic. They include the loss of key talent and valuable corporate memory, higher turnover and absenteeism, decreased motivation and increased stress due to job insecurity, loss of customers, decline in entrepreneurship, innovation and risk-taking, erosion of public relation and brand reputation, increased legal and administrative costs [5].

Downsizing has real potential for reinvigorating an organization and improving its competitiveness, but often it is undertaken as a spectacular quick fix, for example by laying off a big corporate CEO to make the front page of the Wall Street Journal and send highly visible signals to investors. This quick fix may be applauded by the market and provide short term return, but generally fails to produce long-term shareholder value. To be financially effective in the long term and socially responsible, downsizing practices should be conceived and executed as part of a wider transformation effort. They should reflect actions clearly aligned to business vision, mission and strategy. They should be integrated within clearly communicated transformations motivated by continuous strategic improvements, and follow corporate social responsibilities as articulated in the company's "ethics and code of conduct". For example, a TQM culture could implement a continuous effort to optimize staffing requirements (coined *rightsizing* [254]), or a new business model could be designed that communicates give-or-take policies at the onset for prospective employees (e.g. Netflix's grand prize and Uber's X models).

Following a downsizing, the consultant should devise recovery measures. For example, he/she could facilitate organizational changes and redistribution of responsibilities, update in-house skills, develop support structures for employees to revitalize moral, and develop marketing initiatives to improve public reputation.

9.2 Business Strategy

9.2.1 Marketing Strategies

The section of this book on *4P* and *5C* (Chap. 8, Sect. 8.4.4) overviewed some key levers whereto execute marketing strategies. The focus was on framing an exhaustive landscape of implementation levers. In contrast, the current section adopts a high-level narrative about initiating and formulating a marketing strategy, and points to specific issues in the consultant's intervention. The discussion covers the same levers as in Sect. 8.4.4 but adopts an executive style upstream and downstream of these simple *checklists*. To be clear, this is not to say that you wasted your time reading through the primer section! These checklists remain useful structures when thinking about marketing.

Marketing crosses inter-departmental boundaries

It is well known that marketing encompasses market research and promotional activities, less so that it radiates through adjacent activities such as new product concept development, sales and distribution. Because it may affect many business disciplines, top management often experiences difficulty in decision-making when it pertains to marketing initiatives. The decision-making process is a key challenge that consultants should diagnose at the onset [5]. And when helping with implementing a new decision-making process, it is useful to remember a unifying guideline from TQM: quality is a relative term that should revolve around *delivering value to the customer*.

Market feedback is essential

Peter Drucker famously said [46] that *the customer rarely buys what the business thinks it is selling*. But with current IT capabilities, it is becoming possible for a consultant to start the diagnostic by identifying current problems and latent needs in the market [97]. He/she can thereby articulate the solutions that an organization is currently providing, and from these insights go back to align growth objectives with the client [76].

Where should the consultant start?

As a first step, the consultant could analyze the whole product line of his/her client from the perspective of the customers and consumers. In doing so, he/she might quickly spot some underperformers, products or services that have higher cost/sales ratios relative to other products in the pipeline and seem to be delivered more because they fit organization's capabilities than market's demand. At this stage the goal is to recommend strategic additions or deletions in the product line and, eventually, discover unsuspected markets or competitors.

The customer's perspective

The following 4-point guide could be used to frame a marketing consulting project:

Penetration

What customers' problems and circumstances can the client's product solve more efficiently than its competitors? What modifications in the client's offerings would bring an even better solution? In addition to product features and prices, what other dimensions may reflect customers' latent needs? These dimensions are generally: simplicity, convenience and accessibility [19]. When addressing these questions, creativity may be bolstered by mapping out latent needs in the network of diverse stakeholders such as distributors, key influencers, end-consumers [255].

Market development

In addition to current customers, who else presents similar problems/jobs-to-be-done? Can marketing be extended in the global market with little adaptation to local conditions?

Product development
Can the client's facilities and skills be used to offer a new effective and affordable solution?

P & L
Can the new solution be marketed with an acceptable profit?

Challenges in consulting projects
Below is a brief review of some characteristics and challenges of consulting projects aimed at assisting a client with marketing strategies in six common areas: pricing, sales management, customer relationship management, promotion, distribution and product development.

Pricing
In a customer-centered approach, pricing decisions should go beyond simple cost criteria and include insights on market demand. This is sometimes referred to as the *economic value of a product* to a customer, or customer value proposition [76]. The consultant may be brought in to establish new procedures for setting prices, or test and evaluate the effect of price changes with real (*pilot*) market feedbacks. Market feedbacks are essential to effectively implement a pricing strategy given the complex and potentially deleterious (e.g. brand damage) nature of ticking with price elasticity.

Sales Management
The consultant can help improve sales by implementing training and motivation programs for the sales staff. Financial reward-based programs may bolster the incentive of sales staff or channel partners to prioritize the most profitable items, prioritize larger accounts, subscribe new customers, attract potential customers, etc. Another option is to redistribute the sales network according to profitability measured per geographical area, account or channel.

Finally, IT may be leveraged in various ways. IT can make the job of the sales staff more efficient by delivering information to his/her customers and reducing his/her time spent for logistics. IT can also extend the reach of advertising campaigns that may instill curiosity and interest even before a customer gets in contact with the sales staff.

Customer Relationship Management
Attracting new customers and building loyalty is facilitated by IT computer applications that maintain databases of customer habits, preferences and past purchases, build ties between sellers and customers (website, instant messaging, minute information, complementary services) and overall increase customer satisfaction. Nowadays, customer relationship management is essentially an IT activity [256]. Although when it pertains to decision-making it remains a traditional business concept; an *attitude* towards customers [257].

Promotion

The palette of promotional campaigns contains many tools and programs (see *4P* in Chap. 8). The consultant should always optimize the consistency of the promotional campaigns between each other and with the client's vision and objectives. Many authors recommend starting the strategy formulation process with *preset objectives* rather than *allocated resources* [46, 76, 226]. Milan Kubr [5] notes that organizations often set their marketing expenditure as a percentage of sales and this results in initiatives that do not mutually support each other. He advises that the consultant starts with setting objectives, develops a strategy, and if resources are not aligned with requirements from the developed strategy then to set more modest objectives and devise a new strategy, until it fits allocated resources.

A consultant's potential challenge when assisting a client with its advertising campaigns is the sensitive nature of the relationship between the consultant and the client's advertising partners, because both parties have an advisory role vis-à-vis the client.

Finally, it is worth mentioning here again the impact of IT on promotion initiatives, as new technologies afford an opportunity to target individuals or groups within a large population (e.g. customized coupons at CVS).

Distribution

Physical distribution processes have become computerized. In consequence, these processes are more sophisticated, distributors are more concentrated and wholesalers are more bypassed. In turn, demand for this kind of consultancy has declined [5]. The consultant's intervention revolves less around developing channel policies and generally focuses on selecting distribution options (e.g. shall the client bypass wholesalers?) as well as implementing IT management systems.

Product development

The consultant's role in product development is often to ensure reliable and efficient distribution of market information. Key drivers of product development include potential market size, competing products, prices, forecast on competitors response, customer's experience and ability of distributors to provide the right service for the product. The consultant's objective is to ensure that usable data is provided to all functional groups involved (R&D, production, finance, sales and back to marketing), and that all parties are in agreement on market assumptions eventually made. He/she might have to redesign pre-existing capabilities that are used to gather market information, or develop capabilities from scratch.

Packaging may have many objectives, and as such is an intrinsic part of product development. It can enhance customer experience, be used for other purposes (e.g. Nutella containers which are also glasses), get attention of customers, create a coherent brand image (e.g. a matched set of containers for a range of products), protect against damages or simply meet dimension requirement imposed by retailers or transporters.

9.2.2 Small Business Innovation Strategies

Consulting for "small" clients, e.g. startups and subsidiaries, entail many characteristics that differ from assignments with large business clients [258]. First of all, the consultant is often enrolled to assist with asking the *right questions*. He/she may make leaders aware of common issues and best practices, while this type of assistance proves only marginally valuable to larger incumbents because the latter have already developed, tested and refined some core capabilities.

For small organizations, a cheaper, simpler, *one-stop-shop* consulting service tends to be appropriate [5]. Small organizations often need help with preparatory work related to marketing, funding and regulations, in addition to the usual consulting deliverables. The deliverables then tend to take the form of instruments and checklists for self-diagnostic and planning, combined with personalized advices supplementing these instruments. Indeed the consultant may greatly increase client satisfaction by limiting his/her intervention to suggesting appropriate actions and learning packages, stimulating discussions, and reviewing results with the clients, rather than carrying-out the actual task of interacting with potential markets and collaborators or even developing strategies. This provides the client with a less advanced but cheaper and more flexible value proposition, as it covers a broader set of business and regulatory issues within the scope of one single engagement.

Early-stage ventures always face key decisions that pertain to outsourcing vs. insourcing activities [226] as they cannot allocate sufficient time and resources to all operational aspects. For example, bookkeeping and accounting are more efficiently outsourced to external providers.

Entrepreneurs need the help of trusted advisers to avoid the trap of buying unnecessary, unreliable, or excessively complex IT applications and equipment proposed to them by some vendors [69]. The consultant should also help a new business define clear objectives before it invests time and energy developing an online presence. Over the past 15 years, companies of all sizes have rushed to develop a website because it was thought to be essential, but were unable to create and sustain one that served a clear business purpose and was perceived by customers as real help [259].

Some consulting firms are sponsors of startup organizations, through equity-based or partnership agreements (e.g. consulting/coaching role). Incubators for example provide space and facilities for several new businesses clustered under one roof, which in turn facilitates network and access to financial and non-financial supports [260]. Whether to start in isolation or to work with an incubator is in itself a key question that all entrepreneurs should consider. Consultants can help answer this question.

9.3 Corporate Strategy

9.3.1 Resource Allocation Portfolio Strategies

The art of financial decision-making can be distilled into just a few principles, which are discussed in this section.

The corporate portfolio strategies described in the Primer section provide simple roadmaps to avoid cash traps and achieve competitive positioning that results in net cash generation, even when the growth of some businesses slows down. The underlying principle is simple: mature businesses generate more cash that they can reinvest productively, thus the excess cash is best used to support younger growth initiatives. Potential for growth is an asset explicitly accounted for on the stock market. This is reflected by the wide variation of stock prices for large conglomerates that report similar ROI but whose growth potential is perceived differently [261].

Savvy resource allocation decisions traditionally focus on three critical drivers of performance: *competitive advantage*, *return on capital* and *growth* [261]. The underlying principle is again simple: without competitive advantage, returns decrease; without adequate returns, growth destroys value; and without growth projects, strategies à la me-too don't sustain superior performance.

A first guideline to allocate resources toward a sound portfolio objective is to adopt a holistic *top-down* approach to decision making [262]. Capital allocation should not be made at the project level [261], priorities should be set against overall business objectives. One-size-fits all performance metrics (e.g. market share, operating profits) should also be avoided, or at least seconded by tailored metrics. Indeed a project may deliver obvious technical value or immediate financial return but not create much synergy with, or even cannibalize, other concurrent initiatives. And in contrast, a seemingly less attractive project might reveal its upsides only through the lenses of tailored metrics set against overall business objectives [261].

A second guideline is to *categorize* the different business initiatives and opportunities using fact-based metrics that articulate *low-* vs. *high-* vs. *erratic-* return business segments [261]. Low-return initiatives should be further investigated down to root causes and ultimately restructured or divested. High-return and erratic-return businesses, in contrast, should be the objects of growth strategies. The healthiest businesses (the *stars* in the BCG matrix parlance) that bring high return and are in the exponential phase of their life cycle may be allocated substantial amounts of resources without delay, in order to sustain and bolster their currently growing path to success. For segments with less obvious potential, those with high return but low growth (*cash cows*) or with high growth but low return (*question marks*), there exist many subtle opportunities often overlooked or mishandled by market incumbents. For example, the consultant may look at economic linkages between adjacent industrial sectors and consider potentially disruptive innovations that leverage customer *jobs-to-be-done* in term of simplicity, convenience, accessibility, or affordability [5, 46]. In doing so, he/she may articulate *over-served* or *not-yet-served* market spaces [19]. These overlooked opportunities discussed in more details in the section on innovation are, ultimately, what leads to success (rising star) or failure (divestiture).

The programs available to achieve growth include three classic options, namely M&A, partnering, or starting from scratch (internal teams or subsidiaries). Selecting the best option for the portfolio depends on the client's financial capabilities, desire for control and sense of urgency.

9.3.2 Outsourcing Strategies

Outsourcing a set of activities may imply fundamental changes in strategy and organizational design. A consultant should be able to explain the key advantages and disadvantages of outsourcing, and help launch the partnership. Hence, these three aspects are discussed below.

Key advantages of outsourcing include access to new skillsets, freeing up management time to focus on core competencies and reduced capital intensity (transfer from fix to variable costs). Outsourcing to a large provider gives access to economy of scale, leverages brand reputation of the partner and permits staying abreast of fast-changing technologies/innovations.

Potential roadblocks with outsourcing include the increased level of bureaucracy and decreased level of in-house innovation capabilities. This can be particularly daunting in market spaces competing on responsiveness, flexibility and attraction/ retention of human talents.

A consulting assignment on outsourcing may be structured in five phases (Adapted from Ref. [5]):

1. Analyze cost-benefits of outsourcing certain activities by defining the client's core and non-core activities
2. Analyze different types of outsourcing relationships and partner options, addressing the scope of services, business processes, roles and responsibilities, organization and staffing, work plans, deliverables, schedules, budgets, level of control
3. Select a partner, establish lines of report and performance targets
4. Formulate a transition plan for all activities that will be moved from the client organization to the outsourced partner
5. Monitor the plan to evaluate progress and enable the client to make adjustments

9.3.3 Merger and Acquisition Strategies

Mergers and acquisitions are more expensive than simple partnership or outsourcing, and require a higher level of commitment, which in turn incurs a higher level of risk. But it lets an organization quickly expand and reposition, enjoy full control, and eventually become more flexible and competitive in its marketplace.

Unfortunately many M&A initiatives fail to deliver expected results [263], often because the price was too high or because *culture clashes* occur between the two organizations (in term of management style, values, priorities). The phenomenon of culture clash is particularly frequent in M&A between a large conglomerate and a small player [264]. Lesson learned, the consultant should thus inquire and give top executives some ideas of how things look from the perspective of the other organization.

An M&A assignment may be structured in five phases:

1. Articulate capabilities that the client needs in view of its short- and long-term objectives, product line and market demand (*capability gaps*)
2. Analyze cost-benefits of developing the capability internally vs. acquiring it from outside
3. Evaluate different targets by examining financials and strategic fit (*synergies*) with regards to marketing, production capacity, staff capabilities, organizational culture, management style
4. Formulate a transition plan, including lines of report and performance targets
5. Monitor the plan to evaluate progress and enable the client to make adjustments

9.3.4 Collaboration, Cooperation (and Coexistence…) Strategies

Networks are spreading globally [265] as an effective tool to improve companies' productivity by focusing on the things they do best. Networks help companies stay tune to the changing and diverse needs of their customers and rapidly transform their supply, production and distribution systems.

In the high-tech industries (e.g. microprocessors, communication), an emerging phenomenon of *coexistence* between competitors in a same marketplace has been documented [266]. By cooperating, competitors can improve their productivity and competitiveness by easier access to innovation and new technology, and can share risks and liabilities.

The revolution in the economics of information brings the concept of *signaling* to unprecedented new levels. Examples of coexistence between competitors now include *contract manufacturing* between electronic firms, *network intelligence* between IT firms, *network incubator* between startups, partners, advisors and venture capital firms and *virtual continuous teamwork* between partners spread across different geographic locations and different time zones.

Conclusion

The goal of this book was to overview all major aspects of the management consulting industry, including elements that are expected to become core to the value proposition in the near future. It aimed at a "scientific" introduction to management consulting and data science and at discussing the emerging role of information technologies in consulting activities. Again the reader is invited to consider these facts, models, tools (i.e. both elementary and more advanced) and suggestions as a simple aid to thinking about reality, a backbone to facilitate discussion and creativity over concrete issues.

Traditional consulting activities have already been augmented by computer-enabled methodologies, and have already moved from a primarily judgment-based to a more process-based value proposition. Given the recent disruptive changes in the "economics of information" across nearly all industries, it is fair to believe that this trend will continue and that the role played by computational data analysis in management consulting will also increase further.

Because computer-based analytics (a.k.a. *big data*) flourished only 4–5 years ago, covering and blending together the essential "qualitative" concepts of management consulting with the essential "mathematical" concepts of data science has never been attempted in just one book. The intention of this book was clear thus. But the result is incomplete. Even though the bridge between management sciences and data sciences is evolving fast, it is unfinished, and the interested reader should develop it further on its own. There is no standard or best practice for this endeavor just yet, and clearly some links are still missing between data science and management consulting. But again, the goal of this book was a complete overview of management consulting and relevant data science concepts, not a simple toolkit to apply blindly.

The next step for the interested reader thus, is what professional consultants and data scientists are already doing, albeit separately: choose between management consulting or data science, or a specific subject matter, and study this topic in isolation. Then choose another one. This works. Now if this book achieved its goal, what you might have considered highly technical or complex earlier is now part of your comfort zone, or at least out of the unknown. And this will make your next step much easier to undertake.

So for this next step…

J. D. Curuksu, *Data Driven*, Management for Professionals,
https://doi.org/10.1007/978-3-319-70229-2

You will find hundreds of references in the bibliography categorized per chapter, meaning that material on overall industry, IT disruption, client-consultant relationship, data science and strategy for example can be found in Chaps. 1, 2, 4, 6, 7, 8 and 9, respectively.

To be more concise and offer my opinion, below is a quick selection. For further reading on the IT disruption: *Consulting on the Cusp of Disruption* from Clayton Christensen et al. is a visionary article on how computer-based analytics may eventually disrupt the management consulting industry. For further reading on the client-consultant relationship: *The critical success factors in the client-consultant relationship* from Steven Appelbaum is also a visionary article as it first describes success factors from survey results but then leverage the data analytics method of "Machine Learning" to select and prioritize the most relevant factors. For further reading on data science: *Naked Statistics* from Charles Wheelan is one of the best book ever written on statistics because it covers all essentials in data science using an applied, pedagogic approach that anyone who can read can understand. And if you do have a technical background you will still find there a refreshing overview of fundamentals written in an elegant and fun language. Finally, for further reading on strategy and traditional management consulting activities: *Management Consulting* from Milan Kubr is a reference, with 1000 pages that run through the different types of traditional consulting approaches.

The accompanying website econsultingdata.com is a resources platform in management consulting that can be accessed from the MIT consulting club's website and a few other partner websites. It contains original contents as interactive versions of some parts of the book (e.g. industry snapshots of Chap. 1, consulting toolbox of Chap. 3), but also redirects toward books, articles, tutorials, reports and consulting events. It is a simple way to leverage the extensive list of references contained in this book, and to thereby pursue your own literature review.

References

1. Evans C, Holmes L (2013) Re-tayloring management: scientific management a century on. Gower Publishing, Farnham
2. McAfee A (2012) Big data: the management revolution. Harv Bus Rev 90(10):60–68
3. Kiechel IIIW (2010) The lords of strategy. Harvard Business School Press, Boston
4. Christensen CM, Wang D, van Bever D (2013) Consulting on the cusp of disruption. Harv Bus Rev 91(10):106–150
5. Kubr M (2002) Management consulting – a guide to the profession. International Labor Organization, Geneva
6. Graham M (2012) Big data and the end of theory? *The Guardian,* Mar 9
7. French WL, Bell CH (1998) Issues in consultant-client relationships. In: Organizational development, 6th edn. Prentice Hall, Upper Saddle River
8. Canback S (1998) The logic of management consulting (part I). J Manag Consult 10(2):3–11
9. IBISWorld (2016) Global Management Consultants. IBISWorld Inc, Los Angeles
10. Edwards J (2016) Management consulting in the US. IBISWorld Inc, Los Angeles
11. Greiner L, Metzger R (1983) Consulting to management. Prentice-Hall, Englewood Cliffs
12. Edwards J (2014) IT consulting in the US. IBISWorld Inc, Los Angeles
13. Manyika et al (2011) Big data: the next frontier for innovation, competition, and productivity. McKinsey Global Institute, New York
14. Drucker PF (2006) Classic Drucker – the man who invented management. Harvard Business School Press, Boston
15. Stern CW, Deimler MS (2006) The Boston Consulting Group on Strategy. Wiley, New York
16. McCarthy JE (1964) Basic marketing – a managerial approach. Irwin, Homewood
17. Porter ME (1985) Competitive advantage: creating and sustaining superior performance. Simon and Schuster, New York
18. Christensen CM (1997) The innovator's dilemma: when new technologies cause great firms to fail. Harvard Business School Press, Boston
19. Anthony S, Johnson M, Sinfield J, Altman E (2008) The innovator's guide to growth. Harvard Business School Press, Boston
20. Turk S (2014) Global pharmaceuticals and medicine manufacturing. IBISWorld Inc, Los Angeles
21. Turk S (2015) Brand name pharmaceutical manufacturing in the US. IBISWorld Inc, Los Angeles
22. Saftlas H (2013) Healthcare: pharmaceuticals. S&P Capital IQ, New York
23. Lubkeman M et al (2014) The 2013 Biopharma value creators report. The Boston Consulting Group
24. Phillips J (2015) Hospitals in the US. IBISWorld Inc, Los Angeles
25. Phillips J (2015) Specialty Hospitals in the US. IBISWorld Inc, Los Angeles
26. Silver S (2013) Healthcare: facilities. S&P Capital IQ, New York
27. Haider Z (2015) Commercial banking in the US. IBISWorld Inc, Los Angeles

J. D. Curuksu, *Data Driven*, Management for Professionals,
https://doi.org/10.1007/978-3-319-70229-2

28. Hoopes S (2015) Global commercial banks. IBISWorld Inc, Los Angeles
29. Goddard L (2015) Private banking services in the US. IBISWorld Inc, Los Angeles
30. Hoopes S (2015) Investment banking and securities dealing in the US. IBISWorld Inc, Los Angeles
31. Hoopes S (2014) Global investment banking and brokerage. IBISWorld Inc, Los Angeles
32. Hoopes S (2014) Global insurance brokers and agencies. IBISWorld Inc, Los Angeles
33. Alvarez A (2014) Magazine and periodical publishing in the US. IBISWorld Inc, Los Angeles
34. Ulama D (2014) Global Newspaper Publishing. IBISWorld Inc, Los Angeles
35. Ulama D (2015) Global computer hardware manufacturing. IBISWorld Inc, Los Angeles
36. Kahn S (2015) Software publishing in the US. IBISWorld Inc, Los Angeles
37. Kahn S (2015) Telecommunication networking equipment manufacturing in the US. IBISWorld Inc, Los Angeles
38. The Economist (2015) Planet of the phones, vol 414. The Economist, London
39. Brennan J et al (2013) Tough choices for consumer-good companies. McKinsey Global Institute, New York
40. Manyika J (2012) Manufacturing the future: the next era of global growth and innovation. McKinsey Global Institute, New York
41. Crompton J (2015) Global oil and gas exploration and production. IBISWorld Inc, Los Angeles
42. Ulama D (2014) Electric power transmission in the US. IBISWorld Inc, Los Angeles
43. McKenna F (2015) Global airlines. IBISWorld Inc, Los Angeles
44. Soshkin M (2014) Global cargo airlines. IBISWorld Inc, Los Angeles
45. Figures provided by McKinsey representatives at all on-campus information sessions that took place during bi-annual recruiting campaigns over the (2013–2015) academic period at Harvard University and MIT
46. Anthony SD (2012) The little black book of innovation. Harvard Business School Press, Boston
47. Drucker P (1985) Innovation and entrepreneurship. HarperCollins, New York
48. Jespersen L (2009) Client-Consultant Relationships – an analysis of the client role from the client's perspective, MS Thesis, Copenhagen Business School, Copenhagen
49. Deltek (2009) Success factors for consulting firms. Deltek Inc, Herndon
50. AMCF – Association of management consulting firms. www.amcf.org
51. The trend toward accompanying the client in the implementation phase (at least in "initiating" the implementation) is recognized by management consulting firms of diverse sizes across all market places. It was for example emphasized at each Big Three (McKinsey, BCG and Bain) seminar given at Harvard University and MIT in 2013–2015
52. Davenport TH (2013) Analytics 3.0. Harv Bus Rev 91(12):64–72
53. Kiron D, Prentice PK, Ferguson RB (2014) Raising the bar with analytics. MIT Sloan Manag Rev 55(2):29–33
54. White House Press Release (2012) Obama administration unveils big data initiative, Office of Science and Technology Policy, Washington. www.whitehouse.gov
55. Assuncao et al (2014) Big data computing and clouds: trends and future directions. J Parallel Distrib Comput 79:3–15
56. Herman M et al (2013) The field guide to data science. Booz Allen Hamilton, McLean
57. Diment D (2015) Data processing and hosting services in the US. IBISWorld Inc, Los Angeles
58. Kessler S (2015) Computers: commercial services. S&P Capital IQ, New York
59. Sarkar et al (2011) Translational bioinformatics: linking knowledge across biological and clinical realms. J Am Med Inform Assoc 18:354–357
60. Shah S (2012) Good data won't guarantee good decisions. Harv Bus Rev 90(4):23–25
61. Harford T (2014) Big data: are we making a big mistake? *Financial Times*, Mar 28
62. Ratner M (2011) Pfitzer reaches out to academia – again. Nat Biotechnol 29:3–4
63. Think Tank Report (2012) Strategic report for translational systems biology and bioinformatics in the European Union, INBIOMEDvision Inc

64. Lesko L (2012) Drug research and translational bioinformatics. Nature 91:960–962
65. Marx V (2013) The big challenges of big data. Nature 498:255–260
66. Power B (2015) Artificial intelligence is almost ready for business, *Harvard Business Reviews Blog*, Mar 19
67. Proctor L, Kieliszewski C, Hochstein A, Spangler S (2011) Analytical pathway methodology: simplifying business intelligence consulting, In: SRII Global 2011 Annual Conference, IEEE, pp 495–500
68. Nichols W (2014) How big data brings marketing and finance together, *Harvard Business Reviews Blog*, July 17
69. Fogarty D, Bell PC (2014) Should you outsource analytics? MIT Sloan Manag Rev 55(2):41–45
70. Ross JW, Beath CM, Quaadgras A (2013) You may not need big data after all. Harv Bus Rev 91(12):90–98
71. Babej ME (2013) McKinsey & Company: building a legendary reputation through secrecy, *Forbes*, Oct 2
72. Schoemaker PJ (1995) Scenario planning: a tool for strategic thinking. Sloan Manag Rev 36(2):25–50
73. Garvin DA, Levesque LC (2005) A note on scenario planning. Harv Bus Rev 306(3):1–10
74. Porter ME (1996) What is strategy. Harv Bus Rev 74(6):61–78
75. Christensen CM, Raynor ME (2003) The innovator's solution. Harvard Business School Press, Boston
76. Johnson MW (2010) Seizing the white space: business model innovation for growth and renewal. Harvard Business School Press, Boston
77. Kim W, Mauborgne R (2004) Blue ocean strategy. Harv Bus Rev 82(10):76–84
78. Hammer M (1990) Re-engineering work: don't automate, obliterate. Harv Bus Rev 68(4):104–112
79. Stewart T (1993) Re-engineering, the hot new managing tool, *Forbes*, Aug 23
80. Arnheiter ED, Maleyeff J (2005) The integration of lean management and six sigma. TQM Mag 17(1):5–18
81. Davenport TH (2013) Process innovation: re-engineering work through information technology. Harvard Business School Press, Boston
82. Davenport TH (1993) Need radical innovation and continuous improvement? Integrate process re-engineering and total quality management. Plan Rev 21(3):6–12
83. Morgan DL (1997) The focus group guidebook. Sage, London
84. Henderson B (1970) The product portfolio. BCG Perspectives, Boston
85. Kubr M (2002) Management consulting – a guide to the profession. International Labour Organization, Geneva
86. Rigby D (2013) Management tools – an executive's guide. Bain & Company, Boston
87. Crouch S, Housden M (2012) Marketing research for managers. Routledge, New York
88. Shapiro EC, Eccles RG, Soske TL (1993) Consulting: has the solution become part of the problem? MIT Sloan Manag Rev 34(4):89–95
89. Siegel E (2013) Predictive analytics: the power to predict who will click, buy, lie, or die. Wiley, Hoboken
90. Cochran WG (2007) Sampling techniques. Wiley, New York
91. Wheelan C (2013) Naked statistics. Norton, New York
92. Murphy AH, Winkler RL (1984) Probability forecasting in meteorology. J Am Stat Assoc 79(387):489–500
93. Stern P, Schoetti JM (2012) La boîte à outils du Consultant. Dunod, Paris
94. Heller K, Davis JD, Myers RA (1966) The effects of interviewer style in a standardized interview. J Consult Psychol 30(6):501–508
95. Groves RM (2004) Survey errors and survey costs. Wiley, Hoboken
96. Hey T (2010) The big idea: the next scientific revolution. Harv Bus Rev 88(11):56–63
97. McAfee A, Brynjolfsson E (2012) Big data: the management revolution. Harv Bus Rev 90(10):61–68

98. Davenport TH, Patil DJ (2012) Data scientist: the sexiest job of the 21st century. Harv Bus Rev 90(10):70–76
99. Kiron D, Prentice PK, Ferguson RB (2012) Innovating with analytics. MIT Sloan Manag Rev 54(1):47–52
100. Silbiger S (2012) The ten day MBA. HarperCollins, New York
101. Quinn JB (1999) Strategic outsourcing: leveraging knowledge capabilities. MIT Sloan Manag Rev 40(4):9–21
102. Lovallo D, Viguerie P, Uhlaner R, Horn J (2007) Deals without delusions. Harv Bus Rev 85(12):92–99
103. Nikolova N, Devinney T (2012) The nature of client–consultant interaction: acritical review. In: Clark T, Kipping M (eds) The Oxford handbook of management consulting. Oxford University Press, Oxford
104. Drucker PF (1979) Why management consultants? In: Melvin Z, Greenwood RG (eds) In the evolving science of management. Amacom, New York
105. Schein EH (1978) The role of the consultant: content expert or process facilitator? Pers Guid J 56(6):339–343
106. Nees DB, Greiner LE (1986) Seeing behind the look-alike management consultants. Organ Dyn 13(3):68–79
107. Tomenendal M (2007) The consultant-client Interface – a theoretical introduction to the hot spot of management consulting, Institute of Management Berlin, 31. Berlin School of Economics, Berlin
108. Schein EH (1999) Process consultation revisited: building the helping relationship. Addison-Wesley, Reading
109. Tilles S (1961) Understanding the consultant role. Harv Bus Rev 39(6):87–99
110. Shafritz J, Ott J, Jang Y (2015) Classics of organization theory (chapter 5). Cengage Learning, Boston
111. Willems JC, Polderman JW (2013) Introduction to mathematical systems theory: a behavioral approach. Springer, New York, p 1. Introduces the exclusion principle
112. Schön D (1983) The reflective practitioner. How professionals think in action. Basic Books, New York
113. Mulligan J, Barber P (2001) The client-consultant relationship. In: Management consultancy: a handbook for best practice. Kogan, London
114. Cookson P (2009) What would Socrates say. Educ Leadersh 67:8–14
115. Chelliah J, Davis D (2011) What clients really want from management consultants: evidence from Australia. JIMS 6:22–30
116. Alvesson M (1993) Organizations as rhetoric: knowledge-intensive firms and the struggle with ambiguity. J Manag Stud 30(6):997–1015
117. Werr A, Styhre A (2002) Management consultants: friend or foe? Understanding the ambiguous client-consultant relationship, international studies of management. Organization 32(4):43–66
118. Wright C, Kitay J (2002) But does it work? Perceptions of the impact of management consulting. Strateg Chang 11(5):271–278
119. Kieser A (2002) Managers as marionettes? Using fashion theories to explain the success of consultancies. In: Management consulting: emergence and dynamics of a knowledge industry. Oxford University Press, Oxford
120. Clark T (1995) Managing consultants: consultancy as the management of impressions. McGraw Hill, London
121. Pinault L (2000) Consulting demons: inside the unscrupulous world of global corporate consulting. HarperCollins, New York
122. Jackson B (2001) Management gurus and management fashions: a dramatistic inquiry. Routledge, New York
123. Kihn M (2005) House of lies: how management consultants steal your watch and then tell you the time. Warner, New York

124. McDonald D (2013) The firm: the story of mcKinsey and its secret influence on American business. Simon and Schuster, New York
125. Genzlinger N (2013) Tipsy corporate intrigues that fill billable hours: house of Lies, with Don Cheadle and Kristen Bell, *New York Times*, Jan 11
126. Nikolova N, Devinney TM (2009) Influence and power dynamics in client-consultant teams. J Strateg Manag 2(1):31–55
127. Clark T, Salaman G (1998) Creating the right impression: towards a dramaturgy of management consultancy. Serv Ind J 18(1):18–38
128. Turner AN (1982) Consulting is more than giving advice. Harv Bus Rev 60(5):120–129
129. Appelbaum SH, Steed AJ (2005) The critical success factors in the client-consulting relationship. J Manag Dev 24:68–93
130. Tseganoff L (2011) Benefit and best practices of management consulting. IMC Jordan, Jordan
131. Schein EH (1997) The concept of client from a process consultation perspective. J Organ Chang Manag 10(3):202–216
132. Argryis C (1970) Intervention theory and method: a behavioral science view. Addison Wesley, Reading
133. Heller F (2002) What next? More critique of consultants, gurus and managers. In: Critical consulting: new perspectives on the management advice industry. Blackwell, Malden
134. Simon A, Kumar V (2001) Clients' views on strategic capabilities which lead to management consulting success. Manag Decis 39(5):362–372
135. Vogl A (1999) Consultants in their clients'eyes. Across Board 36(8):26–32
136. Fullerton J, West MA (1996) Consultant and client working together? J Manag Psychol 11(6):40–49
137. Bergholz H (1999) Do more than fix my company: client's hidden expectations. J Manag Consult 10(4):29–33
138. AMCF Code of Ethics (2015) Association of Management Consulting Firms, AMCF. www.amcf.org
139. IMC USA Code of Ethics (2012) Institute of management consultants USA, IMC USA. www.imcusa.org
140. McNamara C (2006) Maintaining professionalism. In: Field guide to consulting and organizational development, authenticity consulting. Authenticity Consulting LLC, Minneapolis
141. Glückler J, Armbrüster T (2003) Bridging uncertainty in management consulting: the mechanisms of trust and networked reputation. Organ Stud 24(2):269–297
142. Blacklock C (2015) 15 consulting questions for successful client discovery. www.9lenses.com
143. Redbus2us (2015) What questions are asked to consultants in typical Client interview from Managers? What to respond? www.redbus2us.com
144. Rasiel EM, Friga PN (2002) The McKinsey mind. McGraw-Hill, New York
145. David T (2014) Your elevator pitch needs an elevator pitch, *Harvard Business Reviews Blog*, Dec 30
146. Baldoni J (2014) The Leader's guide to speaking with presence. American Management Association, New York
147. Quoting a McKinsey representative met at an on-campus information session that took place during the 2013 recruiting campaign at Harvard University
148. Cosentino MP (2011) Case in point. Burgee Press, Needham
149. Rochtus T (2012) Case interview success. CreateSpace, Charleston
150. Cheng V (2012) Case interview secrets. Innovation Press, Seattle
151. Thiel P, Masters B (2014) Zero to one: notes on startups, or how to build the future. Crown Business, New York
152. Ries E (2011) The lean startup: how today's entrepreneurs use continuous innovation to create radically successful businesses. Random House, New York
153. Ansoff HI (1957) Strategies for diversification. Harv Bus Rev 35(5):113–124

154. Kohavi R (1995) A study of cross-validation and bootstrap for accuracy estimation and model selection. IJCAI 14(2):1137–1145
155. Lee Rodgers J, Nicewander WA (1988) Thirteen ways to look at the correlation coefficient. Am Stat 42(1):59–66
156. Cover TM, Thomas JA (2012) Elements of information theory. Wiley, New York
157. Kullback S (1959) Information theory and statistics. Wiley, New York
158. Gower JC (1985) Properties of Euclidean and non-Euclidean distance matrices. Linear Algebra Appl 67:81–97
159. Legendre A (1805) Nouvelles méthodes pour la détermination des orbites des comètes. Didot, Paris
160. Ozer DJ (1985) Correlation and the coefficient of determination. Psychol Bull 97(2):307
161. Nagelkerke NJ (1991) A note on a general definition of the coefficient of determination. Biometrika 78(3):691–692
162. Aiken LS, West SG, Reno RR (1991) Multiple regression: testing and interpreting interactions. Sage, London
163. Gibbons MR (1982) Multivariate tests of financial models: a new approach. J Financ Econ 10(1):3–27
164. Berger JO (2013) Statistical decision theory and Bayesian analysis. Springer, New York
165. Ng A (2008) Artificial intelligence and machine learning, online video lecture series. Stanford University, Stanford. www.see.stanford.edu
166. Ott RL, Longnecker M (2001) An introduction to statistical methods and data analysis. Cengage Learning, Belmont
167. Tsitsiklis (2010) Probabilistic systems analysis and applied probability, online video lecture series. MIT, Cambridge. www.ocw.mit.edu/courses/electrical-engineering-and-computer-science/6-041-probabilistic-systems-analysis-and-applied-probability-fall-2010/video-lectures/
168. Nuzzo R (2014) Statistical errors. Nature 506(7487):150–152
169. Goodman SN (1999) Toward evidence-based medical statistics: the p-value fallacy. Ann Intern Med 130(12):995–1004
170. Lyapunov A (1901) Nouvelle forme du théorème sur la limite de probabilité. Mémoires de l'Académie de St-Petersbourg 12
171. Baesens B (2014) Analytics in a big data world: the essential guide to data science and its applications. Wiley, New York
172. Curuksu J (2012) Adaptive conformational sampling based on replicas. J Math Biol 64:917–931
173. Pidd M (1998) Computer simulation in management science. Wiley, Chichester
174. Löytynoja A (2014) Machine learning with Matlab, Nordic Matlab expo 2014. MathWorks, Stockholm. www.mathworks.com/videos/machine-learning-with-matlab-92623.html
175. Becla J, Lim KT, Wang DL (2010) Report from the 3rd workshop on extremely large databases. Data Sci J 8:MR1–MR16
176. Treinen W (2014) Big data value strategic research and innovation agenda. European Commission Press, New York
177. Abdi H, Williams LJ (2010) Principal component analysis, Wiley interdisciplinary reviews. Comput Stat 2(4):433–459
178. Dyke P (2001) An introduction to Laplace transforms and Fourier series. Springer, London
179. Pereyra M, Ward L (2012) Harmonic analysis: from Fourier to wavelets, American Mathematical Society. Institute for Advanced Study, Salt Lake City
180. Aggarwal CC, Reddy CK (2013) Data clustering: algorithms and applications. Taylor and Francis Group, Boca Raton
181. Box G, Jenkins G (1970) Time series analysis: forecasting and control. Holden-Day, San Francisco
182. Peter Ď, Silvia P (2012) ARIMA vs. ARIMAX - Which approach is better to analyze and forecast macroeconomic time series. In: Proceedings of 30th international conference mathematical methods in economics, pp 136–140

183. Chen R, Schulz R, Stephan S (2003) Multiplicative SARIMA models. In: Computer-aided introduction to econometrics. Springer, Berlin, pp 225–254
184. Kuznetsov V (2016) Theory and algorithms for forecasting non-stationary time series, Doctoral dissertation, New York University
185. Wilmott P (2007) Paul Wilmott introduces quantitative finance. Wiley, Chichester
186. Hull JC (2006) Options, futures, and other derivatives. Pearson, Upper Saddle River
187. Torben A, Chung H, Sørensen B (1999) Efficient method of moments estimation of a stochastic volatility model: a Monte Carlo study. J Econ 91:61–87
188. Rubinstein R, Marcus R (1985) Efficiency of multivariate control variates in Monte Carlo simulation. Oper Res 33:661–677
189. Hammersley J, Morton K (1956) A new Monte Carlo technique: antithetic variates. In: Mathematical proceedings of the Cambridge philosophical society, vol 52, pp 449–475
190. Sugita Y, Okamoto Y (1999) Replica-exchange molecular dynamics method for protein folding. Chem Phys Lett 314:141–151
191. Hill TL (1956) Statistical mechanics: principles and selected applications. McGraw Hill, New York
192. Brin M, Stuck G (2002) Introduction to dynamical systems. Cambridge University Press, Cambridge
193. Karplus M, McCammon JA (2002) Molecular dynamics simulations of biomolecules. Nature 9(9):646–652
194. Case D (1994) Normal mode analysis of protein dynamics. Curr Opin Struct Biol 4(2):285–290
195. Alpaydin E (2014) Introduction to machine learning. MIT Press, Boston
196. Matlab R (2015a) documentation. www.mathworks.com/help/stats/supervised-learning-machine-learning-workflow-and-algorithms.html#bswluh9
197. Harrell F (2013) Regression modeling strategies: with applications to linear models, logistic regression, and survival analysis. Springer, New York
198. Breiman L (2001) Random forests. Mach Learn 45(1):5–32
199. Owens J et al (2008) GPU computing. Proc IEEE 96:879–899
200. Simonyan K, Zisserman A (2014) Very deep convolutional networks for large-scale image recognition. arXiv 1409.1556
201. Cho K, Van Merriënboer B, et al (2014) Learning phrase representations using RNN encoder-decoder for statistical machine translation. arXiv 1406.1078
202. Mnih V, Kavukcuoglu K et al (2015) Human-level control through deep reinforcement learning. Nature 518(7540):529–533
203. Kumar A, Irsoy O et al (2016) Ask me anything: dynamic memory networks for natural language processing. In: international conference on machine learning, pp 1378–1387
204. LeCun Y, Bengio Y, Hinton G (2015) Deep learning. Nature 521(7553):436–444
205. Schmidhuber J (2015) Deep learning in neural networks: an overview. Neural Netw 61:85–117
206. Brown C, Davis H (2006) Receiver operating characteristic curves and related decision measures: a tutorial. Chemom Intell Lab Syst 80:24–38
207. Gero J, Udo K (2007) An ontological model of emergent design in software engineering, *ICED'07*
208. Hundt M, Mollin S, Pfenninger S (2017) The changing English language: psycholinguistic perspectives. Cambridge University Press, Cambridge
209. Landauer T (2006) Latent semantic analysis. Wiley, New York
210. Bird S (2006) NLTK: the natural language toolkit. In: Proceedings of the COLING/ACL on Interactive presentation sessions, pp 69–72, Association for Computational Linguistics
211. See for example IBM Watson Alchemy Language API. https://www.ibm.com/watson/developercloud/alchemy-language.html
212. Google's Software Beats Human Go Champion (2016) *Wall Street Journal*, Mar 9
213. Larousserie D, Tual M (2016) Première défaite d'un professionnel du go contre une intelligence artificielle, Le Monde, Jan 27

214. Parkinson's Progression Markers Initiative – PPMI, Michael J. Fox foundation. www.ppmi-info.org
215. Vancil RF, Lorange P (1975) Strategic planning in diversified companies. Harv Bus Rev 53(1):81–90
216. Porter ME (1998) Competitive strategy: techniques for analyzing industries and competitors. Free Press, New York. (original version published in 1980)
217. Wright TP (1936) Factors affecting the cost of airplanes, journal of aeronautical. Science 3(4):122–128
218. Porter ME (2008) The five competitive forces that shape strategy. Harv Bus Rev 86(1):78–93
219. Evans P (1998) How deconstruction drives de-averaging. BCG Perspectives, Boston
220. Evans PB, Wurster TS (1997) Strategy and the new economics of information. Harv Bus Rev 5:71–82
221. Stern CW (1998) The deconstruction of value chains. BCG Perspectives, Boston
222. Perold A (1984) Introduction to portfolio theory. Harv Bus Rev 9:185–066
223. The concept of Strategic Business Unit (SBU) derives from the concept of Profit Center coined by Peter Drucker in the 1940s and was defined by a McKinsey team while consulting for General Electric in the 1970s. This resulted in the GE / McKinsey portfolio matrix and came as a direct response to Bruce Henderson's publication of the growth-share matrix, now referred to as the BCG matrix
224. Rochtus T (2012) Case interview success, 3rd edn. CreateSpace, Charleston
225. Gilbert C, Eyring M, Foster R (2012) Two routes to resilience. Harv Bus Rev 90(12):66–73
226. Johnson M, Christensen C, Kagermann H (2008) Reinventing your business model. Harv Bus Rev 86(12):50–59
227. Kim WC, Mauborgne R (1997) Value innovation: the strategic logic of high growth. Harv Bus Rev 75(1):103–112
228. Rao AR, Bergen ME, Davis S (2000) How to fight a price war. Harv Bus Rev 78(3):207–116
229. Picardo E (2015) Utilizing prisoner's dilemma in business and the economy Investopedia. www.investopedia.com
230. Capraro V (2013) A model of human cooperation in social dilemmas. PLoS One 8(8):e72427
231. Rapoport A, Albert MC (1965) Prisoner's dilemma. University of Michigan Press, Ann Arbor
232. Winger R, Edelman D (1989) Segment-of-one marketing. BCG Perspectives, Boston
233. Bayer J, Taillard M (2013) A new framework for customer segmentation. Harv Bus Rev Digit 6
234. Levitt T (1965) Exploit the product life cycle. Harv Bus Rev 43(11):81–94
235. IBISWorld market research reports are accessible at www.ibisworld.com
236. Humphrey A (2005) SWOT analysis for management consulting, SRI International Alumni Newsletter, 1: 7–8 [this article explains that SWOT originates from some research carried out by A Humphrey at SRI for Fortunes 500 companies in 1960s–1970s – A Humphrey is recognized as the creator of SWOT]
237. Panagiotou G (2003) Bringing SWOT into focus. Bus Strateg Rev 14(2):8–10
238. Hill T, Westbrook R (1997) SWOT analysis: it's time for a product recall. Long Range Plan 30(1):46–52
239. Ries A, Trout J (1981) Positioning: the battle for your mind. McGraw-Hill, New York
240. Lamb CW, Hair JF, McDaniel C (2011) Essentials of marketing. South Western Cengage Learning, Mason
241. Steenburgh T, Avery J (2010) Marketing analysis toolkit: situation analysis, Harv Bus Rev – Case study 510 079
242. Tacit Intellect (2011) PESTLE analysis overview. www.tacitintellect.co.za
243. Tucker FG, Zivan SM, Camp RC (1987) How to measure yourself against the best. Harv Bus Rev 65(1):8–10
244. Craig C, Harris R (1973) Total productivity measurement at the firm level. Sloan Manag Rev 14(3):13–29
245. Martínez-Lorente A, Dewhurst F, Dale B (1998) Total quality management: origins and evolution of the term. TQM Mag 10(5):378–386

246. Feintzeig R (2015) The trouble with grading employees, *Wall Street Journal*, Apr 22
247. Creech B (1994) The five pillars of TQM: how to make Total quality management work for you. Penguin Books, New York
248. ISO Annual Report (2013) www.iso.org/iso/about/annual_reports.htm
249. Poksinska B, Dahlgaard J, Antoni M (2002) The state of ISO 9000 certification: a study of Swedish organizations. TQM Mag 14(5):297–306
250. Hayes RH, Wheelwright SC, Clark KB (1988) Dynamic manufacturing: creating the learning organization. The Free Press, New York
251. Marchand DA, Peppard J (2013) Why IT fumbles analytics. Harv Bus Rev 91(1):105–112
252. Taylor C (2013) A better way to tackle all that data, *Harvard Business Reviews Blog*, Aug 13
253. Cramm S (2010) Does outsourcing destroy IT innovation? *Harvard Business Review Blog*, July 28
254. Hitt MA, Keats BW, Harback HF, Nixon RD (1994) Rightsizing: building and maintaining strategic leadership and long-term competitiveness. Organ Dyn 23(2):18–32
255. Christensen CM, Anthony SD, Berstell G, Nitterhouse D (2007) Finding the right job for your product. MIT Sloan Manag Rev 48(3):38–47
256. Crème de la CRM (2004) *The Economist*, June 17
257. Nguyen B, Mutum DS (2012) A review of customer relationship management: successes, advances, pitfalls and futures. Bus Process Manag J 18(3):400–419
258. Scott M, Bruce R (1987) Five stages of growth in small business. Long Range Plan 20(3):45–52
259. Winter SJ, Saunders C, Hart P (2003) Electronic window dressing: impression management with websites. Eur J Inf Syst 12(4):309–322
260. Bergek A, Norrman C (2008) Incubator best practice: a framework. Technovation 28(1):20–28
261. Hansell G (2005) Advantage, returns, and growth–in that order. BCG Perspectives, Boston
262. Cooper RG, Edgett SJ, Kleinschmidt EJ (2001) Portfolio management for new products. Perseus Books Group, Cambridge
263. Kummer C, Steger U (2008) Why merger and acquisition (M&A) waves reoccur: the vicious circle from pressure to failure, strategic. Manag Rev 2(1):44–63
264. Thomas RJ (2000) Irreconcilable differences. Accenture Outlook 2000:29–35
265. Castells M (2011) The rise of the network society: the information age. Wiley, Malden
266. Zhang H, Shu C, Jiang X, Malter AJ (2011) Innovation in strategic alliances: a knowledge based view. In: Marketing theory and applications, 2011 AMA Winter Educators Conference, pp 132–133

Index

J. D. Curuksu, *Data Driven*, Management for Professionals,
https://doi.org/10.1007/978-3-319-70229-2

Printed by Printforce, the Netherlands

Printed by Printforce, the Netherlands